INTEGRATING CITY PLANNING AND ENVIRONMENTAL IMPROVEMENT

To children and grandchildren everywhere –
who provide a major reason for pursuing
sustainable urban development.

Integrating City Planning and Environmental Improvement

Practicable Strategies for Sustainable Urban Development

Edited by
DONALD MILLER
GERT DE ROO

ASHGATE

Published by
Ashgate Publishing Limited
Gower House
Croft Road
Aldershot
Hants GU11 3HR
England

Ashgate Publishing Company
Suite 420
101 Cherry Street
Burlington, VT 05401-4405
USA

Ashgate website: http://www.ashgate.com

British Library Cataloguing in Publication Data
Integrating city planning and environmental improvement :
    practicable strategies for sustainable urban development. -
    2nd ed. - (Urban planning and environment)
    1. City planning - Environmental aspects 2. Urbanization -
    Environmental aspects 3. Urban renewal - Environmental
    aspects
    I. Miller, Donald II. Roo, Gert de
    711.4'2

Library of Congress Cataloging-in-Publication Data
Integrating city planning and environmental improvement : practicable strategies for
    sustainable urban development / edited by Donald Miller, Gert de Roo. -- 2nd ed.
        p. cm. -- (Urban planning and environment)
    Includes bibliographical references and index.
    ISBN 0-7546-4283-6
    1. City planning--Environmental aspects. 2. Urban ecology. 3. Sustainable development.
    4. Urbanization--Environmental aspects. I. Roo, Gert de. II. Miller, Donald, 1936- III.
    Series.

    HT241.I5825 2004
    307.1'216--dc22

                                                                        2004049050

ISBN 0 7546 4283 6

Printed and bound in Great Britain by MPG Books Ltd, Bodmin, Cornwall

Contents

List of Figures

List of Tables

List of Contributors

M. van den Berg, Professor, Faculty of Spatial Sciences, University of Utrecht, The Netherlands.

W. Biesiot, Faculty of the Department of Energy and Environment, University ofGroningen, The Netherlands.

H. Blanco, Professor, Department of Urban Design and Planning, University of Washington, Seattle, USA.

J. de Boer, Research Co-ordinator, Institute for Environmental Studies, Vrije Universiteit, Amsterdam, The Netherlands.

A. Bus, Faculty, Saxion Polytechnic, Deventer, The Netherlands.

A. Dal Cin, Professor, University of Carlos III, Madrid, Spain.

F.H.J.M. Coenen, Faculty of Public Administration, University of Twente, The Netherlands.

A. Dijkstra, Environmental Planner, Directorate of Public Works, City of Rotterdam, The Netherlands.

J.F. Feenstra, Deputy Head, Institute for Environmental Studies, Vrije Universiteit Amsterdam, The Netherlands.

E. Hoeflaak, Policy Advisor, Environmental Policy Department, Municipality of Rotterdam, The Netherlands.

H.R. Howes, Principal Strategic Planner, Environment Agency for England and Wales, United Kingdom.

A. Kanonier, Faculty of Spatial Planning and Architecture, University of Vienna, Austria.

I. Komoo, Professor and Associate Director, Institute for Environment and Development, Universiti Kebangsaan, Malaysia.

M. de Maaré, Policy Officer, Environmental Agency, Municipality of Groningen, The Netherlands.

D. Miller, Professor, Department of Urban Design and Planning, University of Washington, Seattle, USA.

C. Moller, Senior Urbanist, Municipality of Groningen, The Netherlands.

K.J. Noorman, Department of Energy and Environment, University of Groningen, The Netherlands.

F.H. Oosterhuis, Environmental Economist, Institute for Environmental Studies, Vrije Universiteit Amsterdam, The Netherlands.

J. Oosterveld, Consultant, Lakeshore Planning Group, Toronto, Canada.

M. do Rosario Partidário, Professor, Department of Environmental Sciences and Engineering, New University of Lisbon, Portugal.

J.J. Pereira, Lecturer, Institute of Environment, Universiti Kebangsaan, Malaysia.

R. Piro, Senior Planner, Puget Sound Regional Council, Seattle, USA.

G. de Roo, Professor, Faculty of Spatial Sciences, University of Groningen, The Netherlands.

A.J.M. Schoot Uiterkamp, Professor, Department of Energy and Environment, University of Groningen, The Netherlands.

A. Schreuders, Policy Advisor, Environmental Policy Department, Municipality of Rotterdam, The Netherlands.

M.T.T. Simons, Policy Analyst, Department of Policy Affairs, Ministry of Housing, Spatial Planning and the Environment in The Netherlands, The Hague, The Netherlands.

V.M. Sol, Environmental Chemist, Institute for Environmental Studies, Vrije Universiteit Amsterdam, The Netherlands.

H. Srinivas, Faculty, Department of Social Engineering, Tokyo Institute of Technology, Japan.

J.A. van Staalduine, Deputy Director, Department of Policy Affairs, Ministry of Housing, Spatial Planning and the Environment in The Netherlands, The Hague, The Netherlands.

A. Vaatz, State Minister for the Environment and Regional Development, Sacksen, Germany.

S.D. Vásconez, Senior Consultant, CHRYSALIDA Consultoria Integrada, Quito, Ecuador.

H. Verbruggen, Professor, Institute for Environmental Studies, Vrije Universiteit Amsterdam, The Netherlands.

H. Voogd, Professor, Faculty of Spatial Sciences, University of Groningen, The Netherlands.

E. Zinger, Partner, 'Nieuwe Gracht' Consultancy, Utrecht, The Netherlands.

Preface

As sustainable urban development and management have become increasingly central to governments ranging from local to national in scale, examples of initiatives to implement this goal of sustainability have come to have added significance and utility. This is a principal reason for preparing a new, revised edition of *Integrating City Planning and Environmental Improvement*.

First published in 1999, the 23 cases that this book presents have perhaps even more value today as we move beyond general deliberations concerning the desirability of addressing environmental quality issues, and focus more on strategies to deal operationally with them. The fact that these cases include national, regional, city and neighbourhood programs, results in their providing useful ideas and encouragement to those engaged in planning at each of these levels.

This new, revised edition is intended to be widely available to public officials and planning professionals who are concerned with resolving environmental conflicts through city planning. This edition should also be useful in university courses dealing with this subject, since the cases can both serve as illustrations of specific solutions to problems, and as the basis for critical classroom discussion concerning the appropriateness of these solutions to the specific contexts, and their applicability to other contexts.

This book is one in a series on urban environmental planning, published by Ashgate, that contributes to the mission of the International Urban Planning and Environment Association in fostering the exchange of information concerning practicable ways to resolve urban environmental conflicts. Earlier versions of the contributions included in this volume were presented at one of the biennial symposiums organized by the Association, which received important support from the Dutch Ministry of Housing, Spatial Planning and Environment. We thank Steven van der Veen and Martin Blikman for their suggestions relating the cover design. We greatly appreciate the work of Floris Bruil and Sjoerd Zeelenberg in reformatting the text for this edition, and of Tamara Kaspers and Johan Zwart in redesigning the graphics used in the text.

Donald Miller and Gert de Roo
Seattle, USA and Groningen, The Netherlands

International Urban Planning and Environment Association

Chapter 1

Introduction: Integrating Environmental Quality Improvement and City Planning

D. Miller[1] and G. de Roo[2]

1.1   Introduction

Citizens in cities around the world are expressing greater interest in and are lobbying for improved environmental conditions (Werna and Harpham, 1995). They want clean air, clean water, reduced noise, more vegetation and protection of habitat areas, and safety from dangers from contaminated and unstable soil conditions (Dobson, 1995). These are all seen as contributing not only to their health but to their quality of life (Mega, 1996).

Modern urban planning has addressed some of these concerns since its origins in the early twentieth century. Through the decades, the agenda of planning programs carried out at the local level have added access to sunlight, separation of conflicting land uses, industrial location downwind of residential areas, and traffic control through neighbourhoods. But it was not until the early 1970s that major attention has become focused on environmental conditions by national and local governments (Daly and Cobb, 1989). These initiatives began with environmental legislation establishing programs to improve air and water quality, control noise, and regulate exposure to toxic and carcinogenic chemicals (Chivian et al., 1993). In some instances, these programs have been co-ordinated to streamline regulatory requirements, and to account for the combined effects of several forms of pollution (de Roo, 1993). In most cases, these environmental management programs have taken a sectoral approach, involving separate and unrelated regulation of some of these impacts (Dasgupta et al., 1995).

This book addresses a newly emerging approach by governments to deal with various forms of pollution and threat: the integration of urban physical planning and environmental quality management. While there are as yet no comprehensive models for this integration (Corley et al., 1991), there are numerous cases from around the world that represent progress toward accomplishing this. These innovative programs serve as sources of ideas and provide lessons which can inform similar efforts in other localities. Each of the following chapters report on the experiences of an actual program and on the strengths and shortcomings which have been observed. Most of these contributions are by practising planners and governmental officials who have been close to these programs, thus providing a perspective and sensibility sometimes missing from third-party accounts. The purposes in selecting these contributions are to record and share descriptions of these efforts to integrate urban planning and environmental quality improvement, to provide a critical assessment of these programs based on first-hand experience, and through this to encourage development of similar programs by national and local governments, including further innovation.

Global principles for improving the environmental quality of human settlements in urban areas have resulted from several major international conferences, including the Rio Earth Summit (UN, 1992), and Habitat II in Istanbul (UN, 1996). This book explores a broad set of approaches to implement these policy positions; to make these calls for action operational. Consequently, the presentation of these cases deal not only with the technical aspects of measuring and controlling environmental spillovers, but the institutional, political, and financial aspects of these programs as well.

1.2   Increasing Focus on Environmental Quality

City planning and land use regulation became a widespread responsibility of local governments in many countries in the first decades of the 1900s (Breheny, 1992). The major purposes of these efforts have been to minimise what economists call externalities: those effects by one party which impact another party, usually negatively (Pinch, 1985). Thus the planning and control of urban growth focused on the density and use of land in a manner which would not overtax available infrastructure or create traffic and crowding which would impose on neighbours. In most localities, incompatible land uses such as manufacturing and residential areas were separated in order that the undesirable effects generated by one activity did not burden a more sensitive activity.

This emphasis on orderly urban development was, from the beginning, concerned with avoiding environmental problems in order to improve the quality of life of residents and to protect property values. However, both planning and development controls were for many years narrow in scope (Jennings, 1989). They ignored other issues such as storm water drainage, environmentally sensitive areas, noise, or air and water quality. In large part this was because reliable information concerning the nature and effects of these issues was not available, and responsible governmental agencies did not have staff with the background to apply what was known (Stren et al., 1992). Similarly, there was fear that localities which pursued these issues unilaterally would be at an economic disadvantage in attracting and retaining economic activity, and so such action was politically opposed (Pezzey, 1992).

When acute environmental problems or threats emerged, these were often addressed with specific regulations (Conway and Pretty, 1991). These include protection of water resources, conservation of agricultural and forest lands, siting of sanitary landfills and other facilities which are locationally undesirable, and management of shorelines, floodplains, and wetlands. Few of these specific regulations date further back than about fifty years, and most have not been included as parts of a comprehensive program of planning for the development of cities and urban regions (Ostrom et al., 1993).

Increased public support for addressing a wider range of environmental issues resulted in national water quality and other pollution control acts, and in national environmental policy acts, which began to be adopted in a number of countries in the early 1970s (Douglas, 1983). The broader environmental policy acts require that before any major project or legislation is approved, detailed analysis of its impacts is undertaken, alternatives are considered and comparatively evaluated, and any irreversible and irretrievable commitments of resources which are involved in the proposed action be assessed (White, 1994; Marsh, 1978). This investigation is usually reported in a draft environmental impact statement (EIS), which must be open to review and comment by the public and by other governmental agencies, and these responses considered before a final EIS is prepared and published (McAllister, 1982; Morris and Therivel, 1994). The evidence and findings resulting from this process must be taken into account in reaching a decision on the project or program, regardless of the environmental protection requirements which are included in the approval (Westman, 1985; Cuff, 1994).

In many countries, national governments have required or encouraged local governments to adopt this environmental impact assessment process (Bartone et al., 1994). While this process is usually separate from local plan making and land use regulation, local planning agencies often have responsibility for preparing the EIS for public sector projects and for reviewing the EIS for major private sector development proposals. This has both given rise to developing knowledge about environmental systems and methods for analysing environmental effects, and resulted in planning agencies bringing to their staffs people who have the background to undertake this work. These recent developments facilitate planning programs taking a more active and informed role in dealing with the environmental dimensions of urban growth and change (White, 1992), and position them to integrate these concerns in the preparation of plans which are more comprehensive than they were in the past.

The heightened role of environmental quality as a public policy concern is linked with growing interest in sustainable urban development (Ecologist, 1992; Rees and Roseland, 1991; Mitlin, 1992). Sustainability seeks to guide growth in a manner which does not foreclose options in the long-term future, commonly generations hence (WCED, 1987). It includes economic and social aspects of change in addition to environmental features (Barrow, 1995; Beatley, 1995).

The economic dimension calls for increasing employment opportunities through expansion and attraction of firms which complement rather than have negative implications for social and environmental improvements. The social dimension includes contributing to a sense of community and to social justice among groups within the population. The environmental dimension seeks to conserve bio-diversity for economic, ethical and aesthetic reasons, and to pursue stewardship of environmental services which provide both valuable resources and absorb wastes in a continuing manner (Rees, 1992).

Sustainable development has become one of the major items in planning programs at a variety of stages of economic development (Adams, 1990). Its requirement that long term urban growth balance the three dimensions – economic, social, and environmental – demands knowledge and commitment greater than city planning has involved in the past (Atkinson, 1994). It calls for a systematic treatment of these three dimensions in a manner which we only partly understand: we must supplement scientifically-based approaches with judgement where knowledge is still only partial. While the goal of sustainable development and means to accomplish it are still being explored, this important agenda for local governments is yet another incentive for seeking to integrate the traditional concerns of city planning for the future of the built environment with improvement in our use of natural resources and cleanup of environmental conditions.

1.3   Principles for Integrating Urban Planning and Environmental Quality Improvement

How can we meet the challenge of finding a workable way to integrate physical planning and environmental management? While partial models are being considered and tried, a useful strategy for advancing this development is to begin discussion of a number of principles which can be used to design and evaluate these models. Among these principles are expanded comprehensiveness, developing a reliable and meaningful evidence base, securing broad public participation, including a broad range of alternatives, and carefully balancing objectives. The following elaboration and examination of these principles is intended to initiate this discussion.

Expand Comprehensiveness

For any program to be successful in integrating physical and environmental planning, it needs to include a broad range of the features of urban development. In addition to the traditional concerns of city planning such as density, location and infrastructure, it should address a full range of environmental quality features, but especially those likely to have acute impacts on health and quality of life (Houghton and Hunter, 1994). One tactic is to start with a short list of those environmental features which previously have been covered by legislative programs since these have received some degree of political acceptance. This set of issues can then be expanded, both on the basis of adding those known to have impacts on health and well being, representing irreversible commitment of natural resources, and responding to public expression of interest. In the early stages, a feasible list of environmental issues to be incorporated into planning programs may be limited by staff resources and the ability to communicate a complex set of ideas in a way that is widely understandable. Treating a number of environmental features, along with more familiar physical planning issues, encourages viewing and analysing these effects of urban development in a cumulative manner, and even for the synergistic effects of multiple forms of impact.

Develop a Sound and Appealing Evidence Base

The environmental issues, as with more traditional concerns of urban planning, need to be treated in a convincing manner (Brugman, 1994). This requires that the evidence be reliable, and preferably measurable. The cause and effect relationships of how environmental problems occur, and of their impact on health and well being, are important to keep in mind when designing these measures. Similarly, it is important to make sure that the evidence is valid: that it is a meaningful way to measure the environmental issue of interest. Focusing on outputs rather than inputs commonly contributes to validity. Ideally, the information to be included in this evidence base is already being collected by some agency or organisation, since this will be less expensive than generating new data. Finally, care should be taken in designing or selecting items of evidence so that they are understandable by the public and by decision makers and, where this is not the case, that ways to improve the communication of this evidence are pursued. This may mean that direct measures of environmental features will be preferred, and that indexes which combine these measures be avoided.

As with the principle of comprehensiveness, the most promising planning strategy for designing a sound and appealing evidence base may be to start with a modest set of measures which are then expanded as experience is gained. This first approximation, even if limited, gives people something to review for weaknesses and encourages suggestions for improvements. Such a learning-by-doing process not only stimulates critical reasoning by participants, but increases understanding of the issues and the methods for assessing, as well as a sense of joint authorship of the measures which are adopted. Developing these measures early in the planning process not only will influence the agenda of that process, but will provide a base point for assessing changes which occur as the resulting plan is implemented and increases accountability.

Inclusive Public Involvement

An open participation process is an important feature of initiatives to integrate urban and environmental planning, for several reasons (Leitmann, 1993). Citizens are a source of information and ideas which result from engaging them in designing and pursuing this planning. We do not yet fully understand how to best accomplish this integration, nor have we success-fully sorted out what constitutes desirable physical development and environmental stewardship. Consequently, judgement must be employed to complement what scientifically based evidence we have in reaching decisions on these issues (Barlett, 1990). Matters of judgement depend on a political process for determination, and in a democracy this requires that those affected by the outcomes have a role and voice in determining the results. Since the stakeholders are usually diverse in their values and priorities, such a process must be open to all, and may require efforts to assure that people representing the full range of interests are included in the deliberations.

Because public participation is time consuming and requires resources, governmental officials sometimes would rather avoid opening up the decision-making process to all interested parties, and even feel threatened by the resulting unpredictability which this brings (Mitlin and Thompson, 1995). On the other hand, citizens of cities around the world are increasingly insisting on being involved in these decisions, especially when they will bear the impacts. As a consequence, it is politically wise for officials to embrace a bottom-up approach in developing the purposes and solutions for integrated planning.

A strategy sometimes used is to centralise establishing standards, based on broad public expression of concerns, and then devolve to localities and communities the responsibility for designing the most desired ways to meet these targets. This combination of a bottom-up and a top-down approach is becoming widely adopted by planning programs. When public deliberation takes place in a candid, forthcoming and inclusive manner, it has been referred to as communicative action, and in the most successful situations the planning is based on consensus.

Include Wide Range of Alternatives

Planning involves the deliberate consideration of alternatives in seeking the most desirable course of action to take. The effectiveness of planning is highly dependent on how different the alternatives under consideration are. If the alternatives are modest variations on one another, and thus cluster over only part of the range of possible options, more wide ranging possibilities are ignored by the planning and evaluation process. This is a common weakness in planning practice, and in most cases limits the effectiveness of environmental impact assessments.

To avoid this shortcoming, some planners consciously include unconventional and even controversial alternatives at least during the early stages of the process, in order to stimulate counter suggestions and to provide contrasts in performance as revealed by formal evaluation of the options. Alternatives which are unfeasible or perform poorly with respect to the criteria used will receive low scores in the evaluation phase of the planning process. However even those alternatives which receive low overall priority when assessed may have some comparatively very strong features which can supply ideas for design improvements in those options which evaluation suggests are the most attractive overall.

Experience with least cost planning provides some evidence that this is a means of opening up the range of options considered, including unconventional means of resolving identified problems. This approach also discourages premature closure: that is taking candidate solutions off the table too soon. Devising workable ways to undertake planning which addresses both physical development and environmental quality improvement demands imaginative and even counterintuitive ideas, especially since this is a new kind of endeavour with little precedent.

Balance Objectives

By most definitions, planning is a purposeful activity in that it consciously and explicitly develops a set of objectives and seeks the best means to reach these objectives. Since many objectives are not complementary and often conflict with one another, planning must explore ways to balance these ends in an acceptable manner. Although accomplishing this is a challenge in conventional urban planning practice, adding new environmental objectives increases this challenge (North, 1990).

The issue of determining which objectives are more important than others, and by how much, becomes especially important when alternative plans or program designs are evaluated as a means to inform the decision process. These objectives usually identify qualities sought, such as conservation of habitat and adding family-wage jobs. But they also account for negative qualities such as amount of waste generated and the number of people who are homeless, and the financial costs of public and private sector responses to improve conditions in urban areas.

A common methodology for assessing alternatives is benefit-cost analysis, in which the performance of each alternative is estimated in terms of the monetary value of inputs and results. This provides limited analysis in that all implications must be expressed in market terms, which for a number of important objectives may be very difficult (Coase, 1960), and those features which are not bought and sold in the market may be left out of the analysis (Winpenny, 1991). Also, benefit-cost analysis is only appropriately employed when the alternatives under consideration are different ways of accomplishing the same purposes, not for comparing different kinds of plans or programs. Measuring everything in monetary terms does, however, provide a common basis for comparing performance over several categories of effect, and the results are easily understood, which are advantages of this approach.

Another methodology for evaluation directly employs the objectives as criteria for comparing alternatives (Voogd, 1983). Performance of each alternative is estimated with respect to each of these criteria, the criteria themselves are weighted by stakeholders, and either the weighted performance levels are displayed as a table for discussion or they are combined to provide an overall performance score for each alternative which can be compared with the scores for other alternatives. This approach provides clear evidence concerning the strengths and weaknesses of each alternative which facilitates public critical review and discussion of alternatives. It also employs direct measures of effects and deals with the identity and relative importance of the objectives in an explicit manner. While this multiple criteria method of evaluation avoids the need to price all of the effects included in the analysis and so lends itself to considering a wider range of these effects than does benefit-cost analysis, people participating in weighting objectives to indicate their preferences sometimes have difficulty with this task.

These brief descriptions of two major evaluation methods demonstrate that the process of balancing objectives is difficult and needs to be approached with care. But methodologies such as these do have the benefit of making the implications of choices explicit and debatable, which is greatly preferable to engaging in abstract and often vague discussion of options. Structured analysis of this sort also assists people involved in considering alternatives to deal with the complexity of reaching a decision, even if the analysis stops short of providing overall scores of how well each alternative is likely to perform (Panayatou, 1993).

In addition to comparing options on the basis of their performance in general terms, it is important to consider how different groups within the population will likely be impacted: what are the social justice implications of the alternatives (Holmberg, 1992)? Every decision has these kinds of distribution effects, whether or not they are made an explicit part of the deliberation (Miller, 1985). Social distributional analysis may differentiate affected groups on the basis of where they live, their income or other demographic characteristics such as age (Satterthwaite et al., 1995), and even whether they are current or future residents, including generations hence (Partridge, 1981). The importance of these distributional considerations is suggested by the definition of politics as 'who gets, who pays'.

Finally, the possible conflict between objectives is illustrated by the debate over the term sustainable development, the argument being that increasing jobs is at the expense of environmental quality, and vice versa (Goldin and Winters, 1995). The counter argument is that these are not mutually exclusive, using for example evidence that a number of manufacturing firms have realised economies through reducing their use of resources and production of wastes (Schmidheiny, 1992; Grossman and Krueger, 1995). In addition, localities in many parts of the world have come to realise that efforts to attract firms through lax environmental performance requirements is not a successful long-term strategy: that such firms are foot loose and often are not competitive within their sector in the broader market (Munasinghe, 1993).

Expanded comprehensiveness, developing a reliable and meaningful evidence base, securing broad citizen participation, including a broad range of alternatives, and carefully balancing objectives are a demanding set of principles to apply in designing and evaluating new means of integrating urban planning and environmental quality improvement. Yet each of these principles are important in conventional physical planning, and even more so in approaching the task of combining this with a broader range of environmental concerns. They are presented here to encourage discussion of characteristics which programs to accomplish this integration should have, and as a basis for critically assessing existing programs such as those treated in the balance of this book and briefly described in the next section of this introduction.

1.4   Some Initiatives to Combine Physical Planning and Environmental Quality Improvements

The chapters which follow present a range of programs from around the world which have taken the first steps to integrate physical and environmental planning. Some of these are modest programs to measure and record the spatial extent of environmental impacts through time, and so to raise public awareness of how areas occupied by sensitive urban activities are being affected, and to build a constituency for stronger integrative policy.

Other cases report on innovative programs to reduce multiple forms and sources of pollution at their origin. These initiatives include trading mitigation credits, developing regional performance standards allowing some variation among sources so long as targets are met at the effect-shed level, and methods of compensation to offset negative externalities that remain after feasible environmental quality improvements have taken place.

Some of the most ambitious programs to bring together physical and environmental planning combine impact reduction and control of development to spatially separate and shield sensitive activities from remaining externalities. For example, Dutch integrated environmental zoning, which is discussed in the following two chapters, limits new development of residential and related uses in areas with poor environmental quality after mitigation measures have been taken, and considers relocating existing intrusive or sensitive activities in acute cases (de Roo, 1993; Miller and de Roo, 1996; Miller, 1997). Programs with a similar approach combine improving unstable and flood prone sites and limiting development near them, preserving habitat and open uses through dedication of land as a condition of development, and storm-water utilities which protect streams and increase groundwater recharge where payments by property owners are based on the amount of impervious surface on the site (Stein, 1993).

The organisation of this book consists of five parts. The first four parts consist of descriptive and critical discussions of programs undertaken by various levels of governmental jurisdiction and a range of spatial scales: national, regional, city, and neighbourhood. Cases drawn from each of these spatial scales share a number of characteristics, but there also are a number of institutional, analytical, and political differences between working at one of these scales and another. Effort has been made to include cases from several countries within each of these parts of the book, to provide a basis for comparing the sorts of opportunities and problems confronted in each of these national contexts, and thus to highlight some of the factors which must be taken into account when attempting to transfer program ideas from one context to another.

The fifth section of this book focuses on developing and applying sustainability indicators which combine conventional city planning concerns with environmental quality dimensions. This is a relatively new and increasingly widely adopted strategy used by planning programs to expand their comprehensiveness and to monitor change, as encouraged by the Agenda 21 policy promoted by the Rio Earth Summit (Expert Group, 1994). The chapters making up this section represent efforts in five countries and a range of perspectives both in purposes and applications.

National Policy for Integration

Four chapters in part A address one of the most advanced national policy approaches to foster integrating physical and environmental planning: that of The Netherlands. The first two contributions, by van den Berg, and by van Staalduine and Simons, describe the development of the Dutch experimentation with several kinds of policy and programs, share the experience gained from these experiments, and assess next steps. A major lesson learned from the initial efforts is that standards and a regulatory framework developed by the central government required more flexibility to account for local differences in situation and agenda, and that some degree of local discretion is important in order to achieve results which are both meaningful and politically acceptable.

The second two chapters are commentaries on the first two. Dal Chin offers a Spanish perspective, and Blanco a perspective from the US, on the Dutch experience as reported here. Consequently these first four chapters form a dialogue providing not only a useful description of Dutch initiatives aimed at integration, but also both internal and external critical assessments of these and ideas for further improvements.

In the final contribution to this section, Kanonier discusses Austrian spatial planning law, the responsibilities of each of several levels of government, and kinds of plans and regulations used by each. The contrasts between this and the Dutch approach illuminate each and suggest options, as does the author's treatment of Austrian experiments with encouraging infill development to limit sprawl and to protect important open space in urban regions.

Regional Approaches

The next set of contributions explore efforts to integrate planning and environmental management at the regional level in four countries with quite different institutional structures. In the first chapter of Part B, Piro provides an overview of urban growth management in the US, and a case study of the policy and plan review process of the Puget Sound Regional Council. Of value to other localities is the discussion of criteria used in these reviews, including adequate transportation and other infrastructure, co-ordination across local jurisdictional boundaries, and the effects of plans on air quality and other environmental dimensions.

Dijkstra describes the application of the Dutch ROM (spatial planning and environment region) approach, and its application to the Rijnmond which is the main port area of Rotterdam. Here, environmental impacts from industry and traffic include noise and air pollution. Port expansion, high quality housing, nature reserves, and recreation all need to be accommodated in a manner which balances goals of a stronger economy and a better environment. The methodology used involves disaggregating objectives into desired effects, then evaluating projects in terms of their contributions to these 26 desired results. Its application is illustrated by the analysis of annual progress made by 47 projects, in terms of effects and objectives, providing detail useful in adapting this approach for use in other contexts.

The work of the Environment Agency of South East England is presented in the contribution by Howes. This program controls air, water, and land quality, including brownfield sites, land fills, and extraction of aggregates. In this case, the planning system employs environmental assessment techniques to specify requirements for site enhancement at the development stage.

The fourth case of integration at the regional scale, presented by Pereira and Komoo, deals with a program to manage the physical urban ecosystem in Malaysia which combines geohazard prevention and pollution abatement. Here planning uses mapped geohazards information to avoid development which would be threatened by flooding, landslides, or subsidence. Additionally, air pollution from increases in heavy metals, water pollution, solid waste disposal and the new role of recycling, and control of toxic chemicals and waste are included among the responsibilities of the planning effort.

These four cases, from a range of cultural and institutional contexts, provide both similarities and contrasts in their approaches and in the problems which they have confronted when combining environmental concerns with development planning at the regional level. A number of the analytical techniques are shared, though variations in these also provide options which are worth considering. Additional common themes are the recent increases in the variety and quality of environmental information being used, but also the gaps in important evidence which remain.

Integration at the City Scale

The three chapters in this part of the book report on cases from Germany, The Netherlands, and the United States. The first of these reports on the ecological city model project in the Free State of Saxony, which is intended to be a prototype for the programs in other cities. Vaatz, who is the State Minister for Environment and Development in Saxony, also describes progress with environmental protection programs addressing air and water quality, solid wastes, and fuels for heating in the first seven years of the reunification of this state with the rest of Germany.

Coenen discusses Dutch green plans at the local level, focusing on the coherence between development decisions and municipal environmental plans. This is done by assessing 511 decisions made in the seven largest municipalities in The Netherlands, both before and after environmental policy plans were adopted. A number of his conclusions relate to the Piro contribution in the prior section, including the importance of combining environmental and physical development plans.

Blanco reports on a citizen-based program in Greenpoint-Williamsburg; the most polluted portion of New York City, home to a number of lower income ethnic groups, and the location of many waste recovery and disposal activities. The community used a compensating governmental grant to develop an environmental information system, based on the Dutch integrated environmental zoning methodology (Miller and de Roo, 1996). This information is being used as the basis for prioritising site cleanup, regulating emissions, and controlling additional non-residential development.

Focusing on Integration at the Neighbourhood Level

Five contributions present cases of programs which combine physical planning and environmental quality improvement for portions of cities. These include residential areas, reuse of contaminated sites, renewal of a retail street, and development of an industrial area in an environmentally friendly manner. Methodologies used in each of these cases provide ideas for treating similar concerns in other contexts.

Since much of the housing stock for the next twenty to thirty years already exists, Bus explores ways to undertake environmental renewal of residential areas, focusing on ways to reduce or prevent environmental degradation. He reports on a way to accomplish this, which has been employed in a neighbourhood in Groningen, The Netherlands.

In the second chapter of this section, de Maaré and Zinger report on a project to reuse a centrally located site which was previously used by industry, as a location for 1000 new homes, retail and offices. A major problem confronting this renewal project is significant soil contamination, with remediation being paid for by several parties. By integrating physical and environmental planning, the municipality is able to demonstrate the value added by this remediation, protect ground and surface water quality, and to provide access while avoiding traffic interruption and noise.

Oosterveld assesses a current rehabilitation program for a major commercial street in Toronto, and prescribes a set of additional actions which would reorient the project toward sustainable development and management. The methodology which she presents is aimed at balancing economic and environmental issues.

The chapter by Moller presents an environmental approach to planning for manufacturing areas, and an application of this approach in a prototype project in Groningen. The major issues in this project include transport, economic structure, and landscape and ecology. The framework for addressing these issues was worked out through an open participation process involving representatives from all affected interest groups.

Schreuders and Hoeflaak describe an environmental approach for spatial planning used in Rotterdam, called The Right Place for the Environment. This method distinguishes types of areas based on the quality of public transport and existing environmental quality, which determines the density and type of development that is permitted. Their chapter illustrates the process of applying this method, and how the results are used during the several phases of a planning process.

Using Indicators and Analysis to Achieve Integration

The first four parts of this book consist of chapters presenting cases and the methodologies used for combining environmental quality improvement and physical development planning at the scale of a country, region, city, and neighbourhood. This fifth section focuses on a strategy which is applicable at each of these scales: the development and use of a set of indicators in the planning process. Several of the chapters in this part deal with measures of environmental quality, while others explore the use of sustainability indicators which address social, economic and environmental dimensions of urban development. Designing and applying these indicators can not only inform and influence the development of plans, but also are useful in monitoring the results. These six chapters complement those of the first four sections, and elaborate on the methodologies employed by several of the cases discussed in earlier chapters.

This part of the book is introduced by Miller's exploration of the nature and uses of sustainability indicators. His chapter then presents five cases of programs to specify and use these measures; four by public agencies in the Puget Sound region of Washington State and one by Sustainable Seattle, a non-governmental organisation. Each of these are assessed in terms of their purposes, content, and political standing. The experiences of these programs can be of value to other communities considering a similar initiative.

Vásconez presents the project in Quito, Ecuador to develop a set of environmental indicators through a local participatory process. These indicators are intended to promote public environmental awareness, and in this instance focus on conditions of a contaminated river which flows through the city for the purpose of enlisting public co-operation in improving its condition.

In the third chapter of this section, Sol and colleagues present the bubble concept, an approach used in Amsterdam to measure and manage environmental spillovers. This methodology maps the two-dimensional spatial extent of pollution generated by a cluster of sources, with the third or vertical dimension measuring the extent of impact. Substituting an increase in one form of pollution for reductions in other forms may be permitted, so long as the integrated environmental index for the initial or reference level of environmental quality is not exceeded, or the impacts are reduced. This approach commonly results in lower emissions control costs for the firm, and is found to avoid the problem of slowing economic development which often happens when generic emissions standards are applied. Similarly, planning proposals can be assessed in terms of environmental gains and losses which would result for surrounding residential areas.

Partidário and Voogd develop four models for integrating sectoral planning with a spatial dimension. In addition to these procedural approaches, they develop a substantive approach based on assessing the sensitivity of various resources to contemplated planning actions in terms of use, quality, and availability. This framework, which they applied in Lisbon, Portugal, provides a means of employing indicators and analysis to integrate physical planning and environmental quality improvement.

As Schoot Uiterkamp, Noorman and Biesiot point out, most concern for environmental quality has focused on the production rather than the consumption side of economic activities. Their contribution to this book investigates the household as a consumption unit, and its metabolism. This involves tracking the material cycles and energy flows, using a set of indicators, and seeking ways to reduce these and their impacts on the environment. While their research deals with Dutch households, it is designed to be applied in other contexts as well.

In the final chapter, Srinivas underscores the importance of developing an information system for urban environmental management, and reports on the virtual planning lab which has been developed in Tokyo. This initiative involves using the internet to communicate with citizens and to provide open access to data bases and plans. The on-line database facilitates analysis and critical evaluation, and is supplemented by workshops and training sessions. In this manner, all stakeholders can participate in monitoring changes, assessing options, and communicating preferences and concerns to governmental officials and to each other.

Throughout this book, accounts of these cases contributed by people having direct experience with them provide ideas which can be adapted to other physical, political, and economic contexts. They are also useful in stimulating further imaginative solutions to local problems. In these ways, it is hoped that this book will encourage more governments to consider and explore approaches to integrate planning for improved environmental quality and urban development, through demonstrating not only that it can be done but how it can be done successfully.

Notes

1   Donald Miller is Professor at the Department of Urban Design and Planning, University of Washington, Seattle, USA.

2   Gert de Roo is Professor in Planning at the Faculty of Spatial Sciences, Department of
        Spatial Planning, University of Groningen, Groningen, The Netherlands.

References

Adams, W. (1990) *Green Development: Environment and sustainability in the Third World*,
    Routledge, London.
Atkinson, A. (1994) The Contribution of Cities to Sustainability, *Third World Planning
    Review*, Vol. 16, No. 2, pp. 97-101.
Barlett, R. (1990) Comprehensive Environmental Decision Making: Can It Work?, in N. Vig
    and M. Kraft (eds) *Environmental Policy in the 1990s: Toward a New Agenda*, CQ
    Press, Washington, DC, pp. 1-23.
Barrow, C. (1995) Sustainable Development: Concept, Value and Practice, *Third World
    Planning Review*, Vol. 17, No. 4, pp. 369-386.
Bartone, C., J. Bernstein, J. Leitman and J. Eigen (1994) *Towards Environmental Strategies for
    Cities*, World Bank, Washington, DC.
Beatley, T. (1995) Planning and Sustainability: The elements of a New (Improved) Paradigm?,
    Journal of Planning Literature, Vol. 9, pp. 383-395.
Breheny, M. (ed.) (1992) *Sustainable Development and Urban Form*, Pion, London.
Brugman, J. (1994) Who Can Deliver Sustainability? Municipal Reform and the Sustainable
    Development Mandate, *Third World Planning Review*, Vol. 16, No. 2, pp. 129-145.
Chivian, E., M. McCally, H. Hu and A. Haines (eds) (1993) *Critical Condition: Human health
    and the environment*, MIT Press, Cambridge, MA.
Coase, R. (1960) The Problem of Social Cost, *Journal of Law and Economics*, Vol. 3, No. 1,
    pp. 1-44.
Conway, G., and J. Pretty (1991) *Unwelcome Harvest*, Earthscan, London.
Corley, M., M. Smith and S. Varadurajan (1991) A Network Approach to Enhanced
    Environmental Management, *Project Appraisal*, Vol. 6, No. 2, pp. 66-74.
Cuff, J. (1994) SEA: Evaluating the Policies EIA Cannot Reach, *Town and Country Planning*,
    pp. 45-47.
Daly, H., and J. Cobb Jr. (1989) *For the Common Good: Redirecting the economy toward
    community, the environment, and a sustainable future*, Beacon Press, Boston.
Dasgupta, S., S. Roy and D. Wheeler (1995) *Environmental Regulation and Development: A
    cross-country empirical analysis*, World Bank, Washington, DC.
Dobson, A. (1995) *Green Political Thought*, second edition, Routledge, London.
Douglas, I. (1983) *The Urban Environment*, Edward Arnold, London.
Ecologist, The (1992) *Whose Common Future: Reclaiming the Commons*, Earthscan, London.
Expert Group on the Urban Environment (1994) *European Sustainable Cities, Part 1*,
    European Commission, Brussels.
Goldin, I., and L. Winters (eds) (1995) *The Economics of Sustainable Development*, Cambridge
    University Press, Cambridge.
Grossman, G., and A. Krueger (1995) Economic Growth and the Environment, *Quarterly
    Journal of Economics*, Vol. 110, pp. 353-378.
Holmberg, J. (1992) *Poverty, Environment and Development: Proposals for action*, Swedish
    International Development Authority, Stockholm, Sweden.
Houghton, G., and C. Hunter (1994) *Sustainable Cities*, Jessica Kingsley, London.
Jennings, M. (1989) The Weak Link in Land Use Planning, *Journal of the American Planning
    Association*, Vol. 55, pp. 206-208.

Leitmann, J. (1993) *Rapid Urban Environmental Assessment: Lessons from cities in the Developing World*, World Bank, Washington, DC.

Marsh, W. (1978) *Environmental Analysis for Land Use and Site Planning*, McGraw-Hill, New York, NY.

McAllister, D. (1982) *Evaluation in Environmental Planning*, MIT Press, Cambridge, MA.

Mega, V. (1996) Our City, Our Future: Towards sustainable development in European cities, *Environment and Urbanization*, Vol. 8, No. 1.

Miller, D. (1985) Equity and Efficiency Effects of Investment Decisions, in A. Faludi and H. Voogd (eds) *Evaluation of Complex Policy Problems*, Delftsche Uitgevers Mastschappij, Delft, The Netherlands, pp. 35-50.

Miller, D. (1997) Dutch Integrated Environmental Zoning: A comprehensive program for dealing with urban environmental spillovers, in D. Miller and G. de Roo (eds) *Urban Environmental Planning*, Avebury, Aldershot.

Miller, D., and G. de Roo (1996) Integrated Environmental Zoning; An innovative Dutch approach to measuring and managing environmental spillovers in urban regions, *Journal of the American Planning Association*, Vol. 62, No. 3, Summer, pp. 373-380.

Mitlin, D. (1992) Sustainable Development: A guide to the literature, *Environment and Urbanization*, Vol. 4, No. 1, pp. 111-124.

Mitlin, D., and J. Thompson (1995) Participatory Approaches in Urban Areas: Strengthening civil society or reinforcing the status quo?, *Environment and Urbanization*, Vol. 7, No. 1, pp. 231-250.

Morris, P., and R. Therivel (eds) (1994) *Methods of Environmental Impact Assessment*, UCL Press, London.

Munasinghe, M. (1993) *Environmental Economics and Sustainable Development*, World Bank, Washington, DC.

North, D. (1990) Institutions, *Institutional Change and Economic Performance*, Cambridge University Press, Cambridge.

Ostrom, E., L. Schroeder and S. Wynne (1993) *Institutional Incentives and Sustainable Development: Infrastructure policies in perspective*, Westview Press, Boulder, CO.

Panayatou, T. (1993) *Green Markets: The Economics of Sustainable Development*, ICS Press, San Francisco, CA.

Partridge, E. (ed.) (1981) *Responsibilities to Future Generations: Environmental ethics*, Prometheus, Buffalo, NY.

Pezzey, J. (1992) *Sustainable Development Concepts: An economic analysis*, World Bank, Washington, DC.

Pinch, S. (1985) *Cities and Services – The Geography of Collective Consumption*, Routledge and Kegan Paul, London.

Rees, W. (1992) Ecological Footprints and Appropriated Carrying Capacity, *Environment and Urbanization*, Vol. 4, No. 2, pp. 121-130.

Rees, W., and M. Roseland (1991) Sustainable Communities: Planning for the 21st century, *Plan Canada*, Vol. 31, pp. 15-26.

Roo, G. de (1993) Environmental Zoning: The Dutch struggle towards integration, *European Planning Studies*, No. 3, pp. 367-377.

Satterthwaite, D., R. Hart, C. Levey, D. Mitlin, D. Ross, J. Smit and C. Stephens (1995) *The Environment for Children*, Earthscan, London.

Schmidheiny, S. (1992) *Changing Course: A global business perspective on development and environment*, MIT Press, Cambridge, MA.

Stein, J., (ed.) (1993) *Growth Management: The planning challenge of the 1990s*, Sage, Newbury Park.

Stren, R., R. White and J. Whitney (1992) *Sustainable Cities: Urbanization and the environment in international perspective*, Westview Press, Boulder, CO.

United Nations (1992) *The Rio Declaration on Environment and Development: Report of the United Nations Conference on Environment and Development*, United Nations, New York, NY.

United Nations (1996) *The Habitat Agenda and the Istanbul Declaration*, United Nations, New York, NY.

Voogd, H. (1983) *Multicriteria Evaluation for Urban and Regional Planning*, Pion, London.

WECD – World Commission on Environment and Development (1987) *Our Common Future*, Oxford University Press, Oxford.

Werna, E., and T. Harpham (1995) The Evaluation of Healthy City Projects in Developing Countries, *Habitat International*, Vol. 19, No. 4, pp. 629-641.

Westman, W. (1985) *Ecology, Impact Assessment and Environmental Planning*, John Wiley and Sons, New York, NY.

White, R. (1992) The International Transfer of Urban Technology: Does the North have anything to offer for the global environmental crisis?, *Environment and Urbanization*, Vol. 4, No. 2, pp. 109-120.

White, R. (1994) *Urban Environmental Management: Environmental change and urban design*, John Wiley and Sons, Chichester.

Winpenny, J. (1991) *Values for the Environment: A guide to economic appraisal*, HMSO, London.

Part A
National Policy for Integrating Environmental and Spatial Planning

Chapter 2

Towards Urban Environmental Quality in The Netherlands

M. van den Berg[1]

2.1 Introduction

Cities have a long history and have known glory, decay, growth, change, adjustment and renewal. Nowadays society is working on the sustainable development of cities. Spatial policies in the metropolises in The Netherlands were fundamentally changed by the 1968 Urban Planning Law, stipulating the zoning arrangements for existing and new parts of the cities. Environmental legislation got an enormous boost in the late seventies. Ever since then, cautious steps have been taken towards improving the quality of the urban environment. How far have urban policies come as regards the space and the environment we live in? Is there adjustment and synergy, are there conflicts, and how do politicians and planners resolve them? How do we compare with the rest of the world?

2.2 Urban Planning and the Environment at Cross Purposes

Spatial and environmental policies have a great deal to do with each other, though they also often seem to get in each other's way. As is clear from the legislation, there are important differences in the mentalities behind the two fields. Urban planning legislation was formulated in a period focused on economic growth, consuming space, emancipating workers, and providing them with public housing. The laws on the environment were passed as a reaction to the economic waste of raw materials, the pollution of the air, soil and water, and the effects it had on our health and on nature. The following differences have emerged (Spatial Planning Council/Environmental Management Council, 1996). The main spatial decisions are made at the municipal level, while environmental policies are formulated at the national level. Spatial planning balances the pros and cons, with different quality aims from one case to the next; the central norm is always the same for environmental policies, and there is virtually no balancing to be done. Spatial and environmental planning use totally different concepts and strategies, and barely speak each other's language. Spatial policy is mainly a matter for generalists, and environmental policy for specialists.

During the late 1990s, efforts have been made towards a more integral approach to environmental issues, and there has been a tendency to grant more responsibilities to the lower administrative levels. Urban planning is characterised by quantity thinking, and environmental policies by normative thinking. The combination of spatial quality and environmental goals has led to the conception of the sustainable city. But the road has not been a smooth one.

2.3   Sustainable City

In the 1989 National Environmental Policy Plan, barely any mention was made of the local level. The creation of a compact city requires an environmental quality that involves as few impediments as possible to the use of space. In this connection, the problem of pollution is of major importance: littering, negligence, vandalism and the noise the neighbours make the aim of urban planning is to reinforce the quality of cities! (National Environmental Policy Plan 1989).

The concept of durability or the sustainable city is hard to define. In addition to preserving the ecology, we need to think in terms of the vitality of the economy, safety, and social and cultural aspects. Since everything is in a state of flux, sustainability is not a final goal to be achieved, it is a process of development. A process of change, an evolution with three central aspects: time, space, and management.

Time

The aspect of time implies a link to the needs of various groups of people today, but without limiting future generations in their prospects for tomorrow. An underlying principle here is not to deplete natural resources like space. There is thus no leeway in sustainability thinking for shifting the problems to later.

Space

The spatial aspect pertains to the size of the area, in this case of the city. It is virtually unfeasible to speak of sustainable cities without including the spatial structure of the city itself, and viewing it as part of a much larger region. A few examples to illustrate this: the production of waste in a city is related to the processing outside the city limits, polluted city water affects areas located downstream, and the traffic flows generated by the city pass through an entire region. The use of tropical hardwood even transcends national borders. So shifting a problem elsewhere is another thing that is not feasible in sustainability thinking.

Management

In view of the borders that are crossed and the openness of municipal systems, again and again a link has to be established with the proper management level. Municipal, province and national authorities each play a role in achieving sustainability in the cities. The contributions of private citizens, businessmen and agencies are of decisive importance in this respect. Shifting problems to some other management level is not a viable option in sustainability thinking.

Citizen Responsibility

In sustainability thinking, changes in the management practice and mentality are also important. The individual citizen is prosperous and articulate. The citizen is a full-fledged market party and participates. In this connection, participation does not solely mean having rights, it also means having obligations. Obligations that mean dividing up domestic refuse before discarding it, not littering the street, proper automobile use and so forth. The government is no longer the sole party to determine what is good in the general interest. In this sense, making blueprints is no longer appropriate. They are no longer suited to a society in a state of permanent change. In the years to come, the government will serve far more as an umpire, and will be there to combat excesses. The government should mainly stipulate the conditions and supervise the process. Slowly but surely, the political circuit will have to get used to a new role.

The emphasis is to be more on the organisation of processes, on communication and stimulation. It will mainly be the task of the government to organise the debate and decision-making process and to prevent the discussion from being dominated by one-sided interests. The pursuit of new relations is certain to ensue. Officials and politicians alike will be forced to experiment with how they address environmental and spatial conflicts in the metropolises.

2.4   Experimenting with the Sustainable City

Sustainable city developments should be tested in actual practice. There is ample opportunity to design the development of a sustainable city, as is witnessed by the fact that in The Netherlands, more than €30 billion are invested every year in urban renewal and new construction alone. Widespread experimentation is going on. The experiments are mainly focused on the new construction, and less attention is devoted to the environment that is already there. The experiments involving renovated and new buildings are mainly in the instrumental and organisational fields to keep any harm to the urban environment down to a minimum. The methods can be roughly divided into assessment and solution-generating ones.

Assessment Methods

The current norms serve as the point of departure for the assessment methods, which focus on differences between the interests of various functions.

Method used by the association of Dutch municipalities The aim of this method is to separate areas with functions that can affect each other detrimentally, for example residential and industrial areas. In a systematic fashion, the method provides information about environmental features such as noise, odours, dust and danger, and the distances that should be adhered to between an industrial area and a residential area.

The distances merely serve as a suggestion, calculated as they are based on average industrial plants. The method is simple to use, but not very suitable for complex situations. The method leads to extensive land use, which is quite the opposite of the spatial planning goal.

Integral environmental zoning by the Dutch Ministry of Housing, Spatial Planning and the Environment The aim of this method is to arrive at a lasting equilibrium between ecology aspects and the interests of the city. This method, mainly geared towards complex situations, presents data on various pressures that are being put on the environment. If norms are violated, first source measures have to be taken. It isn't until afterwards that urban planning measures can be taken, such as reorganising plants and moving them. The Western Harbour District in Amsterdam is an example of the use of this method.

Environmental achievement system in Amsterdam The aim of this system is to guarantee a minimum environmental quality in the implementation of the zoning plan. The method is applied at urban districts that are about to be built in order to limit the negative environmental effects. It works with a point system. The environmental achievement system focuses on the total result, so that negative environmental effects can be compensated for in the same district. It is flexible in this respect.

Solution-Generating Methods

Solution-generating methods focus on a specific area, and examine the strain on the ecology there in relation to the vulnerability of the environment. These methods play a strategic role in the formulation of construction- and region-oriented environmental policies.

The bell jar concept in Amsterdam The idea behind the bell jar method is to make spatial choices that have the best yields for the environment as well as the economy. This system charts the stress on the environment in a district in relation to the number of people it affects. By linking the various measures to a budget, insight can be gained into the yields of the various measures.

Environmental matrix in Amsterdam The purpose of this instrument, especially designed for the structure plan level, is to be able to present the environmental ramifications of proposed urban planning measures via a multi-criteria analysis in a relatively comprehensive and rapid manner. By way of the environmental matrix, it is possible to: balance various functions at a certain location against each other; balance various locations for a certain function against each other; indicate how the negative environmental effects of an urban planning measure can be alleviated; stipulate priorities for certain developments.

According to the City and Environment approach of the Ministry of Housing, Spatial Planning and the Environment, there are certain exceptions where it is possible to deviate from the norms, specified by the IEZ. This provides some flexibility to local governments in solving environmental health problems within their jurisdictions.

Before employing these exceptions, though, a number of steps have to be taken. First, a strictly source-oriented policy should be implemented. Then environmental interests have to be integrated into the planning at as early a stage as possible. In addition, as much use as possible has to be made of the leeway in the existing legislation. If all these efforts are in vain, there is the option of deviating from certain norms, provided the resulting harm to the environment is compensated for in quality-of-life terms. There is a preference here for compensation in the same environment category. In conclusion, it can be noted that the City and Environment approach includes assessment as well as solution-generating elements.

2.5   What are the Results?

As regards the environmental policies that have been put into effect, the conclusion can be drawn that despite all the efforts, the strain on the environment in The Netherlands is among the highest in Europe. This is caused by the advancing urbanisation, the small size of the households, and the intensive use of the open areas, especially for farming purposes. In addition, there has been a fall in bio-diversity and a rise in energy consumption and in the pressure on the available space. Lastly, due to the odours, the poor quality of the air, the noise and the restricted amount of space, the cities are becoming less liveable (National Institute for Public Health and the Environment, 1996).

As regards the spatial policies that have been put into effect, the conclusion can be drawn that the spatial quality also leaves something to be desired. There has been a sharp rise in the amount of space needed for each resident, space is being increasingly squandered, the shortage of space in the urban areas is relatively extensive, and accessibility is becoming more and more of a problem. In short, as yet there is not much of a prospect of sustainable spatial and environmental quality. In part, this is due to an inadequate integration of spatial and environmental policies. The city is being used more extensively, and the rising morphological density is diametrically opposed to the environmental quality the city hopes to achieve.

2.6   Integration of Spatial and Environmental Policies

The integration of spatial and environmental policies has yet to receive the attention it deserves. Up to now, it has been more of a case of confrontation than of integration, as is illustrated by the assessment and solution-generating methods referred to above. The Environmental Management Council and the Urban Planning Council have drawn up joint recommendations for the Minister of Housing, Spatial Planning and the Environment about the integration of spatial and environmental policies.

The Councils have concluded that urban planning authorities are now faced with a challenge and will have to develop stricter criteria for spatial quality, that environmental authorities will have to find more options for made-to-measure projects and organising in phases, and that administrative authorities will have to focus more on the scale where they can exercise more control over the quality of the environment.

The Councils recommend spatial interventions that not only yield economic profits but environmental ones as well: they have to be sustainable. Thus, the spatial yields of environmental measures should also be evaluated. Experiments should be stimulated in the integration of spatial and environmental policies. In view of the role of all the levels of government in sustainability thinking, the provinces and municipalities alike need to get into the integration mode. There is a great deal to be learned from the direction the North Holland policies are headed in.

2.7   Provincial Policy Plans

In the province of North Holland, possibilities have been explored for integrating the three strategic provincial plans (Environmental Policy Plan, Water Management Plan and Regional Plan), into one policy plan. For the time being, the strategic plans in the environmental, urban planning, economic and traffic policy fields are to be drawn up according to a fixed framework with a development section, an assessment structure and a programming section.

Development Section

The development concept is viewed as the strategic long-term framework. One aim of the development concept is to enlarge the concentration of construction and intensify the use of the space in cities; the rural area is to be expanded.

Assessment Section

The assessment section is where the construction proposals and activities of private citizens and commercial enterprises are evaluated. A Provincial Environmental Policy Guideline has been stipulated. The Guideline provides insight into the existing plans and norms of the province that are relevant to the environmental policies. The province is also working on a sustainability foundation based on three cornerstones: flows, sites and participants.

Programming Section

The programming section is focused on implementation and consists of all the regional, theme and target group projects that have come to be interrelated.

These then are North Holland's development, assessment and programming sections. For the rest, it is not all that relevant whether or not a comprehensive policy plan is drawn up in the province. It is much more important to eliminate the friction, the competition and the differences between the various policy sectors. In other words, starting from simple forms of co-operation among the policy fields, efforts are being made to work towards the development of joint instruments.

2.8 Sustainability in an International Context

In 1996, the European Union published a report entitled *European Sustainable Cities*. The report described what is meant by sustainable cities from metropolitan regions to small towns. The report made the following recommendations. Work on the further integration of economic, social and environmental policies at all the levels of government. Devote more attention to the sustainability issue in metropolitan regions. Establish better coherence in the policies and implementation of the higher levels of government so that the solution to the sustainability question is not frustrated at the lower levels of government. Provide a good exchange of information so it won't be necessary for everyone to start from the beginning all over again.

The conclusion can be drawn that the international findings largely coincide with the Dutch ones. In fact it might be advisable for The Netherlands to take the European Union's report very seriously indeed.

2.9 Agenda for Urban Issues

As regards the sustainability question in relation to today's urban issues, the following recommendations can be made. Try out various approaches to the sustainability question. Experiment and get experience with implementation projects. Emphasise the operational implementation level rather than all-encompassing strategic plans. Sustainability equals choosing and choosing often also means a profits and losses account rather than an everlasting quest for non-existent win-win situations. Sectoralism is fine, providing it is a first step towards an integral approach. Recognise and acknowledge the responsibilities of regional and local authorities in solving the sustainability question.

More than anything, though, the following is important. A bridge needs to be built between urban planners and environment experts. They should not just want to change each other, they should be willing to join forces for the common interest: sustainable urban quality. For this joint objective, they should be able to transcend their own borders.

The city and the environment are not at cross purposes, they reinforce each other to ultimately create a differentiated, liveable and vital city.

Note

1   Max van den Berg is Professor at the Faculty of Spatial Sciences, University of Utrecht, Utrecht, The Netherlands and Director Spatial Planning at the Province of North Holland.

References

Bartelds, H.G., and G. de Roo (1995) *Dilemma's van de compacte stad, uitdagingen van beleid* (Dilemma's of the compact city, policy challenges), VUGA: The Hague.

European Commission, Expert group on the urban environment (1996) *European Sustainable Cities*, Office for Official Publications of the European Commission.

Hein Struben Advies BV (1997) *Naar een werkstructuur voor omgevingsbeleid, intern advies aan de Directieraad van de provincie Noord-Holland* (Towards a Working Framework for Environmental Policy, Internal Recommendations to the Executive Council of the Province of North Holland) Haarlem, The Netherlands.

Ministry of Housing, Spatial Planning and the Environment (VROM) (1989) *Nationaal Milieubeleidsplan 1989* (National Environmental Policy Plan 1989), The Hague, The Netherlands.

National Institute for Public Health and the Environment (RIVM) (1987) *Zorgen voor morgen* (Concerns for Tomorrow), Samsom H.D. Tjeenk Willink BV, Alphen aan den Rijn, The Netherlands.

National Institute for Public Health and the Environment (RIVM) (1996) *Milieubalans '96* (Environmental Balance '96), Samsom H.D. Tjeenk Willink BV, Alphen aan den Rijn, The Netherlands.

PRO (1993) *Sustainable Urban Development, Research and Experiments*, Delft University Press, Delft, The Netherlands.

Spatial Planning Council/Environmental Management Council (RARO/RMB) (1996) *Duurzaam en leefbaar* (Sustainable and Liveable), The Hague, The Netherlands.

UNDP/UNCHS/World Bank (1994) *Land Use Considerations in Urban Environmental Management*, Urban Management Programme, Washington, DC.

Chapter 3

Environment and Space: Towards More Cohesion in Environmental and Spatial Policy[1]

J.A. van Staalduine[2] and M.T.T. Simons[3]

3.1 Introduction

From the end of the 1980s, especially, there has been a growing realisation of the interdependence of spatial and environmental policy. The Dutch government published its viewpoint on the cohesion between these two policy fields at the end of 1997 in a policy document entitled *Environment and Space*. This policy document will focus primarily on these two policy fields; the relationships with other policies concerned with the physical environment such as water, nature, housing and transport, will only be indirectly referred to. The extra value of a broader integration of policies for the physical environment has been acknowledged and will be explored in the future, but it is too complex to be contemplated at present.

The discussion points and development directions outlined in this chapter are for a large part based on the findings of the working group that is preparing the policy document on *Environment and Space*.

3.2 Quality of the Environment

The *Environmental Report* of 1996 stated that in the case of the substances causing the greatest environmental burden, with respect to the period 1985-1995 an absolute reduction in emission levels had been achieved (RIVM, 1996). Over the country as a whole disturbance in general has been reduced largely as a result of a reduction in disturbance from noise and odours. Survey statistics show that of those affected by traffic and industrial disturbance declined from 50 per cent in 1990 to 44 per cent in 1995. In addition the disturbance from odours emanating from agriculture and industry has been reduced from 23 per cent to 18 per cent. Regional and local disturbance percentages vary from the national averages. The (cumulative) noise disturbance caused by industry, and motorway, rail and air traffic is in the *Randstad* and other urban areas higher than in the more rural areas. Dwellings with a noise level of more than the accepted limit of 50 dB(A) are concentrated in the urban areas of the Randstad and parts of the provinces Noord-Brabant and Limburg. Local variations can also be observed within the urban areas themselves.

Dutch norms for the various noise sources are comparable with those in other European countries. The percentage of the Dutch population exposed to noise levels of more than 50 dB(A) caused by road, rail and air traffic and industry is virtually the same as in Germany, France, the United Kingdom and Denmark.

For the rural areas there are strong regional variations in the burden from acidification, eutrophication and groundwater depletion. In contrast to disturbance, where there is a trend towards a national reduction – despite a continuing concentration in urban areas – limits set with respect to these three types of environmental damage have been exceeded in 1994 with respect to 1989 across the country as a whole, especially in southern and central Netherlands (RIVM, 1996). Calculations show that the areas with the highest risks for vegetation are largely the same as the areas where the limits for acidification, eutrophication and groundwater depletion have been exceeded the most. The risks for vegetation are greatest in parts of Noord-Brabant, the edge of the Veluwe, and the Achterhoek.

In the international context there are great variations. The nitrogen damage to the soil in The Netherlands from animal manure and artificial fertiliser is, for example, higher than in any other country in Europe. The great variations in sensitivity and environmental pressure make area-specific solutions in tackling environmental problems extremely desirable. This is also true for disturbance problems; in the urban areas in particular the disturbance problem is a persistent one. The complaints about odour and noise in the Rijnmond area and around Schiphol airport have merely increased in number.

Another observation, again according to the Environmental Report, is that the good results obtained in limiting environmental damage in the future threaten to be cancelled out by the growth in volume of the population, the economy and mobility. Through savings, for example, the energy intensity was reduced by around 0.8 per cent in 1995. This reduction was, however, more than cancelled out by a structural shift of 2.1 per cent whereby, on balance, an increase of 1.3 per cent took place.

3.3   Province and State

This chapter principally addresses the central government approach to the further integration of environment and space. At other levels too – local authorities, water boards and provinces – a great deal is being done to integrate spatial planning (and water management) with environmental policy into a policy for the physical environment. At present the discussions at the provincial level are focused on planning. The province is obliged to draw up a spatial plan, an environmental policy plan, and also a water management plan. Separate production of these plans, which as regards content are closely related, costs a great deal of professional and political effort, time, and money. Each province has looked for methods to better integrate the various types of plan and policy. In this respect a distinction can be made between: co-ordination of procedures, co-ordination of content of policy, and integration of plans and policy.

There are no particular problems in administrative law with respect to the co-ordination steps. More important are the problems such as cultural divisions and differences which have grown historically. A future integration of plans and policy demands a new orientation on the planning systems and legal instruments. In addition one comes up against fundamental questions such as the present and future implementation of policy.

At the national level efforts are being made to break down the divisions between ministries and parts of ministries. Joint approaches to the physical environment are being adopted and the consequences of a stronger integration for the overall philosophy of central government and planning systems are being considered. In the political context the Minister of Housing, Spatial Planning and the Environment has expressed the wish to, in any case, integrate spatial planning and environmental policy and to start the preparations for a national environment and spatial planning policy document in which the existing national policy documents on spatial planning and the environment are incorporated.

3.4   Motivation

Three main reasons can be distinguished for seeking a synergy between environmental and spatial policy.

First, both policy areas need each other. At the local and regional levels only an integrated approach can achieve an improvement in the environment. An attractive design that at certain locations creates irresponsible safety risks is not achievable. Obsolete industrial areas are not restructured, partly because of high land decontamination costs, with the result that expanding private enterprise may be forced to emigrate. At the national level the same problem is apparent, but with another dimension: environmental policy needs spatial planning in order to tackle the various types of pollution. Together they can confront the growth in volume. The reduction in the environmental burden per firm and household threatens to be cancelled out by the growth in volume. Spatial planning also has a role to play here. Through growth in volume, concepts such as the compact city seem to be reaching their limits.

Second, improvement in the implementation of spatial and environmental policy: experiences in the implementation of spatial and environmental policy indicate that benefits can only be achieved through good co-operation. Dutch examples of this are the Spatial Planning and Environment Projects[4] and the City and Environment experiments. Can lessons be learned from these projects that are more widely applicable? Co-operation is, however, no panacea to all the problems in environmental and spatial policy. The application of legal instruments, for instance, needs a great deal of improvement; in infrastructure plans the environmental consequences are considered only after the design has been made, whilst it is also conceivable that environmental considerations, like technical demands, be integrated in the design process.

Third, take local and regional circumstances more into consideration: Dutch environmental policy was, up to and including the First National Environmental Policy

Plan, strongly oriented on achieving generic goals, for which norms were developed which were applicable across the whole country. In areas under heavy pressure the deposition norm can only be met by additional, area-specific measures. Also the generic approach was not always appropriate for vulnerable areas (nature areas, drinking water areas): the norm may be met, but the environmental problem has not been solved. In those cases too, extra measures are required.

3.5   Vision on the Present Relationship between Spatial and Environmental Policy

Environmental policy and spatial policy are both oriented on the quality of the environment (Table 3.1). Each is concerned with different aspects of quality.

   The division of tasks outlined above is becoming increasingly a caricature, however. More and more attention is being devoted within spatial policy to environmental quality and more stringent norms are being set. Within environmental policy specific situations and qualities are being taken more into consideration. It is clear that an improvement in the quality of the environment can be achieved through different approaches – preservation, development, prevention and restructuring – each of which has a separate objective and can make its own contribution to the whole. There is no hierarchy; they complement each other. In spatial and environmental policy terms, on the basis of these approaches different contributions are made to the quality of the environment.

3.6   Environment, Space: Developments over Time

Sustainable development has two sides: preventing the shift of blame for pollution (in space, in time, and to other environmental compartments) and providing development opportunities.

   The first aspect is historically the primary responsibility of environmental policy whilst spatial planning traditionally takes care of the second aspect. Both are, however, inseparably related to each other, as is apparent in the statements of the following major goals.

   Sustainable development is 'a development which provides for the needs of the present generation without at the same time endangering the opportunities of future generations to also provide for their needs' (VROM, 1993).

   Spatial planning has its main objective the promotion of such spatial and ecological conditions that: the real ambitions of individuals and groups in society are realised; and the diversity, cohesion and sustainability of the physical environment are safeguarded as much as possible (VROM, 1990).

Table 3.1   **Differences in approach between environmental policy and spatial planning in The Netherlands**

		Spatial planning	Environmental policy
goal		spatial quality (through weighting of claims)	environmental quality (through demands / norms)
nature of laws		formal	material
plans	national (state)	indicative	guiding
	regional (province)	guiding framework	guiding
	local (municipality)	binding	optional and non-binding
implementation		assessment of initiatives of citizens	through government and target groups and through enforcement
vertical relationship between levels of government		emphasis on 'bottom-up' approach	emphasis on 'top-down' approach
position with respect to policy subject		creating conditions	laying down conditions
weighting of interests		integrated according to functions / areas on behalf of function allocation	integrated according to environmental components and target groups on behalf of quality goals
participation		for everyone	emphasis on advice and consultation
grant of permit		by municipality, via building and planning permits	by central government, province and municipality (various types of permits)
enforcement		by municipality	by central government, province, municipality, etc.

Source: Mastop and van Damme (1996)

The 1970s

In the 1970s it became clear that spatial policy was insufficiently able to protect the quality of the environment adequately. Contamination of land and water and air pollution spread far across geographical boundaries. Influence on behaviour via land use was only possible to a limited extent. An example of this was intensive animal husbandry. Although this was regarded as undesirable in certain areas, spatial planning did not succeed in controlling the growth in the number of pigs, never mind limiting the damage from manure to the land and the water supply.

Environment became an independent policy field and developed its own policy attitude and legal instruments. This often led to conflicts between developers, whose instincts were for optimal solutions, and highly motivated 'managers' who had a range of effective legal instruments at their disposal.

The 1980s

During the 1980s this tension produced initially only ideas about collaboration but in the end led to mutual recognition and a limited form of 'teamwork'. In the Fourth National Policy Document on Spatial Planning, three ways were indicated by which spatial and environmental policy could support each other (VROM, 1990).

Firstly, environmental policy set constraints on land use planning. An example of this is environmental zoning: by creating distance, risks are diminished and disturbance is limited.

Secondly, spatial policy itself contributes to environmental quality. Examples of this are: the concentration of urbanisation, location policy for firms and services, the promotion of public transport, the concentration of waste treatment, the diversion of water flows from other areas, the retention of water of local origin and the development of strategic water supplies.

Thirdly, environmental policy enables spatial developments to be possible for instance by cleaning up contaminated land, by reducing noise in areas where investment is a priority (urban nodal points), and by improving water quality for the benefit of recreation.

The 1990s

The co-ordination between the two policy fields has mainly been oriented in particular on the prevention of 'disturbance' and on the mutual 'acceptance' of goals and concepts. Environmental policy-makers remained managers, planners remained developers.

In the 1990s the following phase has become apparent: the stereotypical division of tasks is being broken down. Within spatial policy attention for environmental quality is growing and more stringent norms are being set, in which regional variations are being taken into consideration. Within environmental policy there is a growing appreciation of specific situations and qualities.

Looking Ahead

In order to give further impetus to the 'relationship', boundaries within both policy areas have to be crossed. We must strive for 'process integration': thinking about and working on each other's primary subject areas from the start of development, preservation, restructuring or prevention with respect to damage to the environment. Environmental management supplies ideas for spatial development. In other words, in football terms, the defender goes forward, provides crosses and even scores. Spatial planning supplies ideas for environmental management. The attacker helps out his defence. End result: total football.

3.7   Three Converging Dimensions

The traditional distinction between environment and spatial planning is beginning to become blurred. Spatial policy is becoming more oriented to management issues, whilst environmental policy is becoming more development-oriented. Management goals appear to be achievable when they can be related to spatial dynamics.

Something similar is evident with respect to the management of material and energy flows. Traditionally this is the domain of environmental policy: the physical environment as a collection of supplies which are connected to each other by chains and flows. In contrast spatial planners are used to perceiving the environment as a collection of functions, connected by networks. In practice both approaches complement each other. Many flows follow natural or human networks. As these networks attract new flows they have a great influence on the location of spatial functions. 'Sight locations' along motorways are, for example, attractive to firms.

A similar shift can be seen in the traditional preferences for particular scales. Environment was always strong at the international and national levels whilst spatial policy possessed its strongest legal instruments at the local and regional levels (such as the local land use plan). In environmental policy, however, more and more attention is being paid to local variations, and in spatial policy the emphasis is shifting in planning from the local land use plan to the supra-municipal level.

3.8   Recent Developments

It is not sufficient to establish that the content of spatial planning and the environment are (again) growing closer together. What are the actual developments underlying this trend and what is the long-term perspective of further integration of these two policy fields in the Dutch context? In terms of actual developments it is important to recognise that in the second half of the nineties a broad policy discussion has been conducted about making (financial) instruments for the environment more compatible and about the need for an integrated weighting of different interests in order to achieve an optimal environmental quality. We shall go into both these aspects.

Financial Flows

Sectors of government policy which comprise the spectrum of environmental and spatial policy (housing, environment, nature, water, traffic and transport) are responsible for a large number of financial flows. This diversity leads to problems in the municipal and provincial authorities who have to elaborate and implement national policy. There are three fundamental causes of these problems.

First, variations in s*etting priorities and programming*: projects are not given the same priority by each successive government. The result is that resources are available too late or are not available at all; the possibilities of 'linking' are reduced (for example simultaneous utilisation of land reconstruction measures and measures to counteract groundwater depletion).

Second, *policy scope* in regulations is limited. Through co-operation between authorities and other actors resources can be deployed more efficiently, but many regulations offer insufficient possibilities for utilising public funds in a wider sense than the specific intention of the regulation itself. It is also often impossible to reserve public funds for a certain period. All kinds of specific conditions are attached. In the calculation of central government's contribution to land costs for housing, for instance, reconstruction costs have to be considered separately. For land decontamination there is a separate financial regime.

Third, differences of an *administrative nature*. Every regulation has its own application procedures, target groups, budgeting pre-conditions (for example, advance payments or not), project and/or programme financing, accountability etc. These differences lead to a great deal of administrative red tape.

In urban areas there are strong moves to introduce more legal instruments for integrated weighting in the interests of the physical environment. The Minister of Housing, Spatial Planning and Environment has announced, following consultation with the regional authorities, that the hitherto separate financial regimes for land costs in housing construction and land decontamination will be combined. The idea is that the regional authorities are granted a certain sum per dwelling on the condition that a number of criteria are met (number of dwellings, proportion to be built in existing urban areas and new locations etc).

Further integration of budgets can be achieved by the setting up of an urban renewal fund, managed by the municipalities. Not only funds for housing construction and land decontamination but also funds for insulation (to counteract noise disturbance), traffic and transport measures, and for sufficient green space amenities could be included in such a fund. The disadvantages of such a fund are, however, also considerable. The greatest disadvantage is that scarce resources are divided up and in that way their efficient use is made more difficult. In addition it would no longer be possible to set and justify national priorities for specific goals.

Not much progress has been made with the amalgamation of funds for rural areas. Partly as a result of the numerous regional divisions indicated in all the national policy documents (each self-respecting government sector has compiled a map of its regional configurations, and in many cases combined with a regulation for financial subsidy) the number of financial flows is large. There is no question of an integrated fund for the rural areas. In 1996, however, several subsidies for the rural areas have been, from the environmental viewpoint, combined into a subsidy for area-specific environmental policy. Separate finance for soil protection areas or tranquillity areas, provided by central government, is therefore a thing of the past. Provinces submit a request to central government every year for this subsidy.

The Ministry of Housing, Spatial Planning and the Environment and the Ministry of Agriculture, Nature and Fisheries agreed in 1997 that a plan of action must be elaborated in order to harmonise central government regulations with respect to the rural areas. The intention is to set up a national 'office window' where regional authorities can submit their requests. Assessment within central government can then take place in an integrated way, recognising the separate policy intentions of the various government sectors.

Integrated Approach

The second important development at the national level is that a large number of projects have been started up that all have the following characteristic: although starting from a sectoral basis only an integrated approach can provide a satisfactory solution to the problems. We refer to three of these projects: the City and Environment Project; the renewed soil decontamination policy; and the modernisation of the Noise Disturbance Act.

City and Environment Project In 1996 the Minister of Housing, Spatial Planning and the Environment informed parliament that research into the relationship between the environment and spatial planning in the urban areas had revealed that many problems exist.

The majority of these problems appears to originate from poor co-ordination at the local level between environmental and spatial planning instruments and, related to this, inadequate co-operation between various departments, as well as from a lack of knowledge about the possibilities within the present regulations to arrive at responsible choices. At the same time it was established that even with optimal co-ordination and communication, situations remain compelling choices that do not produce an optimal environmental quality. Such situations will only increase in number in the future as a consequence of the increasing demands on the limited available space in the urban areas.

The Minister of Housing, Spatial Planning and the Environment has also announced that the opportunity will be given in a limited number of cases to experiment with planning methods, comprising three steps. The first step involves integrating environmental interests at the local level as early as possible in the planning process. The second step is carefully utilising the scope available within the existing regulations. Finally, the third step is departing, under certain conditions, from the national norms and procedures if it is apparent that by following the first two steps no success has been achieved in terms of optimal environmental quality. Environmental loss, which will take place in such a situation, must be compensated preferably within the compartment in which the loss is evident. If that is not possible compensation must be found within other environmental compartments or even outside the environmental domain altogether.

Renewal of policy for soil decontamination The development of strategic projects and housing locations in urban areas is oriented on the integrated improvement of the environment. Strong signals are coming from practitioners that a number of projects are stagnating as a result of land contamination. The cause of this stagnation is the enormous extent of the land contamination, the related drain on the financial resources required for decontamination (many tens of billions of euros), the procedures needed to determine how the costs must be borne and the requirement for multifunctional decontamination. This last point means that the policy intention is to carry out the land decontamination in such a way that all functions will be possible on the decontaminated land.

In view of the enormous extent of the land decontamination operation and the limited national resources, priorities inevitably have to be set. Often land decontamination in urban areas is not so environmentally urgent that it appears high on the list of priorities. Decontamination is, however, necessary if new projects or housing construction are to be carried out. Otherwise an area can remain untouched with all the associated disadvantages, including even sometimes deterioration.

At the national level new policies are being considered to break this deadlock. It has been established that a new balance must be found between policy objectives and the extent of the land contamination. *Towards cleaner land in the long term* is now the starting point; ensure that the land is not contaminated even more, that the contamination be properly documented and tackled step by step. Furthermore, the approach to the land decontamination does not always have to be multifunctional.

The challenge lies in the integration of the land decontamination in an environmentally hygienic way with other processes. It is apparent at the moment, from practical evidence, that (inner city) development locations with little or no land contamination are extremely scarce (the better locations have already been used). An improved implementation of land decontamination is, therefore, more necessary than ever. This improvement is to be found in an accelerated development of cheaper decontamination techniques, the simplification of procedures surrounding land decontamination and more scope for the executive authorities (municipalities, provinces) in tackling the decontamination. This encourages an integrated approach and can accelerate the spatial development. Consideration can be given, for example, to reserve the limited national budget for cases with the greatest environmental urgency whilst with respect to land decontamination in urban renewal, the costs could be borne by a combination of land taxes and possible (additional) subsidies.

Modernisation of the Noise Disturbance Act A similar development can be observed here. In The Netherlands there are stringent noise disturbance norms with respect to road and rail traffic and in the vicinity of industry and airports. In practice, however, it was apparent that in a number of cases, in particular in existing urban areas, these norms could not be enforced. A noise disturbance norm for an area such as the Leidseplein in Amsterdam, where hotels, cafes and restaurants are concentrated, has to be interpreted differently from that in another area or street. A large number of exceptions had to be made, therefore, and exemptions from central government regulations granted. In order to give the municipalities the opportunity to arrive at an integrated assessment in achieving an optimal environmental quality, a national Advisory Committee suggested that the municipalities be given the freedom to determine their own noise levels for certain urban areas. The national government has adopted these suggestions in principle and is in the process of elaborating them in close consultation with regional and local authorities and social organisations.

This advice stated that policy on noise disturbance as a result of road traffic (not motorway traffic) can be decentralised. A legal indicative norm (50 dB(A)) was sufficient, this norm indicating at what noise level there is virtually no risk to public health and to which the municipality can refer if it has no policy of its own. Municipalities receive the powers to set their own norms on condition that they can explain any departure from the indicative norm in their own policies. Municipal noise policy covers all aspects of noise policy, that is to say tackling the problem at source, prevention and removal. This policy is set out in a municipal document (the *noise policy document*), which should be integrated into the land use plans. A stringent demand remains in place, however, that the noise level within the home should not rise above 35 dB(A).

3.9   From Convergence to Perspective

Spatial and environmental policy are growing closer together. That is borne out in practice. It has already been stated that both policy fields are coming closer together via the policy attitudes, the interaction between the policy objects (areas on the one hand and flows on the other) and the scales used. What is now the perspective for the future?

We would formulate the answer to this question as follows: Bring together management and development! The reasoning here is that (spatial) developments offer the possibilities of bringing management goals within reach in an efficient way. Substantial changes are after all easiest to bring about at the moment that a link is made to spatial dynamics. Guidance is only possible if there is movement. On the other hand, developments need a certain base in order to take place at all. The management-oriented policy approach must provide support for desired developments towards environment and space of adequate quality.

An example of an elaboration at the local scale is the design for 'ecological' neighbourhoods: if the environmental department is involved from the beginning, due consideration can be given to separation of waste, district heating, discouraging or excluding car use, use of solar energy via the orientation of dwellings etc.

At the national level consideration needs to be given as early as possible to environmental demands which can play a role in decisions about the construction of infrastructure: which route crosses the least vulnerable or valuable areas, where can noise disturbance cause the least damage, how can the influences on ground-water flows be limited? These and other considerations should play a role before the first designs reach the drawing board.

Last but not least, international agreements are relevant in this context. The location of nuclear power stations in border areas, drinking water extraction with transfrontier implications, the flooding problems of the River Meuse, the development of a European network of high speed trains and also of nature areas are some examples in which more cohesion between management and development would be desirable.

3.10   Consequences for the Spatial Planning System

A new style for a cohesive spatial and environmental policy must start from concrete problems and be given shape by co-operation. The *content* must be 'made-to-measure steering'; in addition to the classical hierarchical steering network, steering and self-regulation can play a role. The key to providing 'tailor-made' solutions lies more in a flexible approach than in a continuous refinement of instruments. With respect to the spatial planning system we must confront the question whether types of plans in the past were not too legal. Driessen distinguishes three options (Driessen, 1995). Integrated planning can take place by co-ordinating existing plans properly; through drawing up one integrated environmental and spatial plan; or by carrying out the integration primarily in the project decision phase on the basis of a 'vision'.

In the design of *legal instruments*, policy-oriented flexibility is achieved by devoting more attention to the various functions of norms in the policy process. In complex situations norms which set limits often provide insufficient flexibility and norms can be better utilised in an indicative way. Furthermore a clearer distinction is desired between political-strategic visions – programmes of content on the basis of which politicians and other actors determine their positions – and legal regulations: rules for the management and preparation of projects.

In terms of the *organisational form*, a clear division of responsibilities between administrative levels is important. Government must keep asking itself whether something needs to be steered, and if so at which scale. Individual authorities must have more possibilities to be able to pursue their own environmental and spatial policy. More attention is needed for project co-ordination and less for plan co-ordination. Environmental and spatial issues do not limit themselves to ministerial competence's and administrative straitjackets, nor even to the boundary between the public and private sectors.

3.11   Consequences for the Framework of Concepts: Re-framing

In order to give teeth to the above perspective for the improvement of the cohesion between spatial and environmental policy, modification of, for example, goals and instruments is necessary. The question is whether this will appear to be sufficient to actually achieve an integrated environmental and spatial policy. In spatial and environmental policy the quality concepts utilised are so different that a really common and cohesive explanation of a concept such as 'environmental and spatial quality' is impossible. A common framework of concepts can mean that issues which require an integrated approach are not regarded primarily from a sectoral viewpoint but from the viewpoint of the quality of the environment. The argument to regard and define the land use, functional and management issues with respect to the physical environment systematically as environmental and spatial issues has been referred to in this context by Mastop as 're-framing' (Mastop and van Damme, 1996). The conscious use of these generic concepts is more than just a trick: it does more justice to the increasing acknowledgement and recognition of the complexity of the issues in our environment and the acknowledgement that technical and legal approaches alone do not bring about the solutions.

3.12   Conclusions

This chapter finishes with the text of an advice from the Advisory Councils for Spatial Planning and the Environment at the request of the Minister with regard to the preparations of the integrated policy document on environmental and spatial planning (RARO, 1996):

> Integration means the removal of all steps which lead to friction, conflicts of competence's and divergence between policy fields. That is to say from simple forms of co-operation between the two policy fields 'at the basis' to, if necessary, the development of joint instruments.

The essence of the discussion on integration can be formulated as follows, according to the Advisory Councils: the challenge for spatial policy is to develop harder, better quantifiable criteria for spatial quality; environmental policy must search for the possibilities for more 'tailor-made' solutions, more scope for negotiation and weighting; and political accountability must relate better to the scale at which the environmental and spatial quality can be 'steered'.

What is needed is a common language, a series of concepts whereby the separate qualities of environmental and spatial policy can be better grasped for policy as a whole. Politicians, administrators, planners and citizens must be able to communicate with each other about quality objectives of locations as a whole but also the individual qualities. The series of concepts must be sufficiently related to present spatial and environmental policy so that it can co-ordinate spatial and environmental goals.

In spatial policy, there is experience with weighting processes with regard to spatial quality. In environmental policy, experience has been built up with supplies, flows and chain management. A combination of these experiences should lead to an adequate framework to achieve a sustainable environmental and spatial quality.

Notes

1   The authors would like to thank Dave Hardy and Wil Akkies for their contribution to this chapter.

2   Mr. J.A. van Staalduine is Deputy Director of the Department of Policy Affairs, Directorate General of the Environment at the Ministry of Housing, Spatial Planning and the Environment in The Netherlands. Van Staalduine is especially involved with integrating spatial planning into the environment.
3   Mr. M.T.T. Simons is employed at the Department of Policy Affairs, Directorate General of the Environment at the Ministry of Housing, Spatial Planning and the Environment in The Netherlands. His responsibilities include integrated regional policy and the Dutch national policy document on environment and spatial policies.
4   These are Regional Projects in which different authorities, departments, and citizens co-operate to solve regional spatial and environmental problems in an integrated way.

References

Driessen, P.P.J. (1995) *ROM-gebieden (ROM-areas)*, Stedebouw en Volkshuisvesting, Vol. 76, Nr. 1/2, pp. 9-21, The Hague, The Netherlands.
Mastop, J.M., and L. van Damme (1996) *Integratie als opgave: overwegingen bij geïntegreerd omgevingsbeleid [Integration as a task: considerations to integrated area-oriented policy]*, Katholieke Universiteit Nijmegen, Nijmegen, The Netherlands.
RARO (Raad voor de Ruimtelijke ordening en Raad voor de Milieuhygiëne) (1996) *Duurzaam en Leefbaar; over de onderlinge afstemming van ruimtelijk beleid en milieubeleid [Sustainable and liveable; about mutual attuning of spatial policy and environmental policy]*, Advice from the Advisory Councils for Spatial Planning and the Environment. RMB 96-03, The Hague, The Netherlands.
RIVM (Rijksinstituut voor Volksgezondheid en Milieuhygiëne) (1996) *The Environmental Report of 1996*, Bilthoven, The Netherlands.
VROM [Dutch Ministry of Housing, Spatial Planning and the Environment] (1990) *Fourth Dutch National Policy Document on Spatial Planning*, The Hague, The Netherlands.
VROM [Dutch Ministry of Housing, Spatial Planning and the Environment] (1993) *Second Dutch National Environmental Policy Plan*, The Hague, The Netherlands.

Chapter 4

A United States Perspective on the Dutch Government's Approach to Seeking Greater Cohesion in Environmental and Spatial Policy

H. Blanco[1]

4.1   Commendable General Approach

The general approach and direction of the Dutch government (Chapter 3; Miller and de Roo, 1997; VROM, 1997) in seeking broader integration of environmental policy and spatial planning is commendable. The United States is far from such an approach. The only attempts to develop a national policy to deal with both spatial planning and environmental protection date to the late 1960s and they were defeated by the mid 1970s (Mandelker, 1976). However, the National Environmental Protection Act, which became effective in 1970, created the federal Environmental Protection Agency.

This act also required an Environmental Impact Assessment (EIA) of all projects that used federal funding. States subsequently adopted similar policies, and the practice of requiring EIAs for significant land use projects is now widespread at the local and state level in the US. This is the chief instrument through which spatial planning and environmental concerns are integrated throughout the United States. As the vast literature on EIA makes clear, there are many problems with relying on the EIA process for an integration of environmental policy and spatial planning, chief among them: its piece-meal nature, focus on consequences and not systems and cycles, its development orientation, its advisory nature.

Beginning in the 1970s, the federal government also enacted policies to regulate water, air, soils, toxic substances, and other aspects of the natural environment. These policies rely on standards or control of specific pollutants. Although state and local governments in the US carry out such policies, environmental policy is primarily determined at the federal level. In contrast, federal and state governments have traditionally had very little control over spatial planning or land use in America. Although emanating from the states, the power to regulate land use and to provide for public infrastructure still lies primarily at the local level.

In the US, the closest we have come to integrating spatial planning and environmental policy is through the state-wide efforts at growth management. Ten states in the US had in 1997 statewide spatial planning in place. Growth management efforts incorporate spatial planning, environmental policies, and infrastructure planning and finance. These programs include various intergovernmental structures and processes to co-ordinate planning and implementation policies across multiple government layers, which may be useful for VROM to examine. Among the state growth management efforts, the closest approximation to the Dutch attempt to integrate environmental and land use planning is new legislation passed in Washington state in 1995 (ESHB 1724) the environmental and land use permitting act (Settle et al., 1995).

Since the Washington state growth management program is one of the strongest in the country, this reform bill is very significant, since it calls for the integration of comprehensive planning and environmental impact review. It reinforces planning by mandating that permitting agencies rely on plans already in place when they review projects; at the same time, the act streamlines the permitting process. Although this act may be moving in the right direction, by more closely linking spatial planning and environmental permit review, it is still far from an attempt to integrate the interests and perspectives of environmental policy and spatial planning. More significantly, no studies have assessed the effectiveness of state growth management efforts (the oldest of which, Hawaii, Vermont, Oregon, and Florida, have been in place since the early 1970s) to deal with regional environmental problems, or compared these efforts with the effectiveness of federal regulations alone.

The US context, even in growth management states, is also plagued by conflicting financial flows at every level of government. Signs of hope are the Intermodal Surface Transportation Efficiency Act (ISTEA) of 1991 whose policy is to develop a transportation system that is 'economically efficient and environmentally sound' and the Clean Air Act Amendments of 1990 which authorise sanctions for failure to meet reduction targets in urban smog that include loss of highway funding (Braum, 1994; Bryner, 1995). These two acts are so interlinked that they have the potential to reshape the collaboration between transportation, land use planning, and air quality sectors.

4.2   Enormity of Environmental Problems

Some of the new policy directions of the Dutch Ministry of the Environment and Housing (VROM) are motivated by the enormity of the environmental problems found, and the lack of economic resources to implement the optimal strategies. For example, with Integrated Environmental Zoning (IEZ), the costs make prohibitive the relocation of whole neighbourhoods or the closing down of factories that provide local employment (Miller and de Roo, 1997); soil contamination is so extensive that many tens of billions of euros would be required for decontamination; or noise disturbance in central city locations violate stated standards constantly, so that the standards cannot be enforced.

VROM's move to provide more flexibility and decision-making power at the local level makes a lot of sense. People who live with environmental harms should have a greater say in how to deal with problems, especially if there are no public resources to remove or significantly reduce the harms or risks. However, if the range of solutions that the localities can resort to are less than optimal, then the turn to greater flexibility and more local control, can be interpreted as a disingenuous attempt by the national government to abdicate responsibility. If the problems are serious and the range of solutions available now do not significantly reduce them, then a two-phased approach may be more appropriate. In the first phase, the local level would be provided more flexibility and power to deal with these problems on a short to medium range time frame. For those problems where the harms or risks remain, VROM could retain the responsibility to address these problems on a longer-term basis, so that as funds become available, these harms are mitigated or removed. VROM could develop a priority list of projects that are acted on as funds become available.

4.3   On Changes Considered for IEZ

Some of the changes proposed to make IEZ more flexible are accepted practice in the United States. There is no national noise code in the US, and so local communities develop their own codes, with variations among localities. Also, within a city, for example, differences are drawn among different residential zones. For example, in New York City (NYC), the Noise Code sets out different noise standards for 3 noise quality zones, and differentiates between lowest density residential districts, higher density districts, and commercial and industrial areas (NYC Department of Environmental Protection, 1992).

Compensation of areas with a high concentration of environmental harms or nuisances is something we have not had experience with in the US. Because of our reliance on EIAs and their piecemeal nature, as well as our reliance on the judicial system to settle land use and environmental controversies, communities are often compensated for the impacts of a particular project through mitigation measures or through amenities in lieu of mitigations. A good example of this is the Environmental Benefits Program in Brooklyn, New York City, which I discuss elsewhere in this volume. This program was created by the city's local environmental agency in response to a court agreement with the State of New York resulting from violations of the Clean Water Act by the local sewage treatment plant. The city set aside $850,000 to provide funds for a series of environmental programs for the community that bore the largest impacts of the environmental violations.

The idea, however, of compensating areas where there is a high concentration of environmental harms from multiple sources, as indicated by the IEZ classification system, is commendable. Industrial and other facilities that pollute or pose a risk to health or safety, either benefit private interests and/or the public at large. These benefits are captured by either private firms, or are spread throughout the population. These facilities, however, have localised impacts. People who reside within the area affected by these environmental harms are bearing a disproportionate impact or cost. If these impacts cannot be eliminated or reduced in a significant way, then the local community should be compensated. In addition to the issue of fairness, compensatory measures serve to internalise the costs of pollution and nuisance, and thus move towards a true cost approach where both the private and public sectors are given more appropriate economic signals.

4.4   On Devolving Decision-Making to the Local Level

I am in great sympathy with the Spatial Planning and Environment (ROM) experiments, the new procedural approach, as reported by Miller and de Roo (1997) which promise a more co-operative consensus-building approach. I believe in strong democracy, of devolving appropriate decision-making power to local and, even, neighbourhood councils. But my notion of democracy is post-environmentalist, that is, I would want to ensure that the decisions/outcomes arrived at do safeguard health, safety, and natural resources. And so my notion of democracy emphasises the need for responsible participation (Blanco, 1994). People entrusted with public decisions should have the knowledge commensurate with the issues involved. If VROM is serious about pursuing a ROM approach, then public education on spatial and environmental planning is crucial.

Also, and as important, local decisions on spatial/environmental issues have regional impacts, and so local decisions should be made within the context of a regional planning process. I understand that the ROM experiments begin within such a regional context, and, thus, some of these comments may be unnecessary. But if we are serious about both democratic process and ecology, we have to contend with the fact that some local decisions require co-ordination at the regional, and some even at the national and the international level. Effective and democratic decision-making processes to deal with environmental/spatial issues require intergovernmental structures and processes. The issue is not so much the state versus the local government, because both have a role to play, but more importantly, the challenge is to identify the issues where regional and other jurisdictional concerns come into play, to figure out the appropriate roles for the various players, and to arrange forums to facilitate such decisions.

In general, it is hard to find fault with VROM's attempt to become more flexible. However, quite often more flexible approaches tend to require more resources or co-ordination. For example, van Staalduine and Simons identify three options for integrated planning: the co-ordination of plans by different state agencies or levels of government; the development of one integrated plan; and integration achieved at the project level on the basis of a 'vision'. But project level integration can only be achieved if the major players involved are, indeed, guided by a common and integrated vision. This way of achieving integration is thus the most challenging, for it requires that officials involved are educated to share this vision and that the public be equally educated to ensure local consensus and support for such decisions.

4.5   From Convergence to an Integrated Perspective

The shift VROM envisages for the future – from convergence to an integrated perspective is very encouraging. The new policy calls for a reframing of the concepts so that 'area-specific environmental management' and 'flow-oriented development' is achieved. This is the logical next step for VROM, and one which I heartily endorse. But I wonder whether VROM recognises what a radical proposition this is. And, if so, how does VROM propose to achieve this transformation in public officials or the public at large? I would argue that such an integration requires adequate investment in retraining the professionals involved as well as in reforming higher education, at least. Government plans that call for a more integrated or systems approach to policy, such as the Planning-Programming-Budgeting Systems instituted by the US federal government in the mid 1960s which sought to rationalise the US budget process, fail miserably, to a large extent because little thought is given to the retraining requirements for creating such radical changes in perspective.

In addition, environmental policy and spatial planning schools are still turning out the same divergent professionals. As van Staalduine and Simons (Chapter 3) note, the knowledge to reframe the concepts is not yet available. This has become painfully evident to me and my colleagues as we work on just such a synthesis, attempting to develop a graduate program on urban ecological planning at the University of Washington. We are in the process, but it is a difficult process, of making the connections between the traditional ecological approach of flows and chains of energy and materials and spatial planning concerns. What is clear is that if we are serious about an integrated outlook on environmental and spatial issues, it will require reform of the professional education of environmental managers, spatial planners, transportation planners, economists, and other professions. This is something that no experimental program in integrated planning in one or two universities can accomplish. I hope the Dutch government recognises the need to invest in research and curriculum development to bring this about for the professions involved. In the US, this need is beginning to be recognised. For example, in 1997, the National Science Foundation has recognised this need by finally opening up the competition for long-term ecological research centre support to urban sites that address social, economic and behavioural issues, and also by setting up a program to support *Integrative Graduate Education and Research Training* (National Science Foundation, 1997). All such programs, however, provide insufficient funding for very few institutions. Widespread, long-term change is not likely to come from such efforts. The Dutch government has the opportunity to lead the world in creating an integrated framework that is taught and reinforced in the training of all the relevant professions.

Furthermore, if VROM is also serious about seeking greater local and public participation in an integrated planning process, then these ideas need also to reach secondary and university education for all citizens. To ensure acceptable environmental outcomes, public involvement requires an informed, knowledgeable citizenry.

4.6   Public Health belongs in the Integrated View

I understand that VROM's environmental policy is driven by public health concerns, and this is probably the reason that public health is not included in the call for reframing the issues. Public health issues were seen as a given, driving the environmental policies. Or perhaps, public health issues were interpreted narrowly, under the conventional medical model. However, under the new conception of health, such as advocated by WHO's Healthy Cities program, health concerns are seen more broadly as part of developing and promoting a healthy life (Davies and Kelley, 1993). If public health issues are interpreted in this broad way, and incorporated into the new integrated approach, both spatial and environmental policy issues can become more immediately relevant to the ordinary person. In addition, incorporating public health within a reframing of the issues, also makes available public health measures as potential strategies, including, environmental health education, or enhanced public health services in areas of heavier environmental loads.

4.7   Conclusions

Dutch government policy seeking to integrate environmental and spatial policy continues to lead the world. US policy in this direction pales in comparison. However, some of the new Dutch initiatives are problematic, and others are so ambitious, they require radical, systemic changes. The devolution of decision-making power to the local level prompted by the enormity of environmental problems can be interpreted as a weakening of resolve to deal with environmental problems. This would be clearly the case, if local governments are not provided or equipped with the means – both resources and legal powers to deal with these issues. Devolution itself is not the right concept. Environmental problems are systemic, and often have regional, national and transnational impacts. They require well-articulated intergovernmental systems to address them. Institutional co-ordination, both horizontal and vertical, will be most important for success.

The type of new perspective that the Dutch government envisages requires radical changes in the entire educational system, from elementary schools through the university level. A wide range of professionals need to be retrained. Greater public participation at the local level also requires a population with almost as much sophistication in outlook as the new professionals. If the Dutch are serious, this new policy direction calls for a quantum change in both commitment and redirection of resources. But, if the Dutch can realise this challenging vision, the whole world will be in their debt.

Note

1   Hilda Blanco is Professor at the Department of Urban Design and Planning, University of Washington, Seattle, USA.

References

Blanco, H. (1994) *How to Think About Social Problems: American pragmatism and the idea of planning*, Greenwood Press, Westport, CT.

Braum, P. (1994) *ISTEA's Planner's Workbook*, The Surface Transportation Policy Project, Washington, DC.

Bryner, G.C. (1995) *Blue Skies, Green Politics: The Clean Air Act of 1990 and its implementation*, CQ Press, Washington, DC.

Davies, J.K., and M.P. Kelley (eds) (1993) *Healthy Cities: Research and practice*, Routledge, London.

Mandelker, D. (1976) *Environmental and Land Controls Legislation*, Bobbs-Merrill, Indianapolis, IN.

National Science Foundation (1997) Integrative graduate education and research training program: program announcement, NSF, Washington, DC.

NYC Department of Environmental Protection (1992), *Noise Code*, New York, NY.

Miller, D., and G. de Roo (1997) Transitions in Dutch Environmental Planning, *Environment and Planning B: Planning and Design*, Vol. 24, pp. 427-436.

Settle, R.L., J.T. Washburn and C.R. Wolfe (1995) Washington State Regulatory Reform: Jekyll and Hyde '95, *Land Use Law*, September 1995.

Chapter 5

Commentary on 'Spatial and Environmental Integration'

A. Dal Cin[1]

> Nothing is good or bad because of the laws. The quality criteria depends on its accommodation of nature, its reason and its usefulness for society. Averroes (Córdoba, XII century)

> Nothing is truth or lie. Everything looks like the colour of the glass you are looking through. Calderón de la Barca (Madrid, XVII century)

5.1 'State of Facts'

It is difficult to give some comments on another's work, especially when the subject is so complex, and able to be seen from so many different viewpoints, as the one we are facing in this debate. Before coming to the matter itself, I want to point out some concepts as frameworks. I call these a 'state of facts'.

Good Urban Quality of Life

Planning in urban areas is facing a big challenge: increasingly bad conditions of life, decreasing budgets versus growing needs, global economic forces guiding the markets, aggressive city competitiveness, segregation, unemployment, marginalisation, etc.

One of the highlighted concepts arising in terms of how to achieve good urban quality of life is innovation. This means especially 'ad hoc' answers to as many generic as specific problems. These answers have to work with a legal framework, with an administrative structure, with a historical background and with an identity based on cultural roots. Of course innovation tries to obtain the maximum profit from the existing frameworks in order to solve the problems quite soon without shocking and shaking changes.

Increasing Role of the Local Level

Local authorities and communities play an important role in the process of achieving good urban quality of life. It is as much a matter of individual as of collective responsibility. The higher governmental levels are so distant from the day to day work that they are not able to produce significant synergies among the measures needed to move towards the aims. Even when we talk about global aims we know that they need effort at the local level to become a reality.

Dichotomy of Spatial Planning-Environment

The field of planning has changed and enlarged during the 20th century. Land use, socio-economic issues, housing, health, heritage, mobility, telecommunication, etc. were concepts included by planning. In the last years the term 'spatial planning' achieves more weight and gives a more comprehensive character to the process.

Another field has appeared in a parallel way: the environment. Environmental problems were faced outside of planning, following other methods. It is the first time that two nearest and interrelated elements, space and environment, are treated as diverse elements. Even if certain administrations tried to avoid the division, creating national or regional structures linking both fields, planning and its implementation still work on their own, assuming certain environmental guidelines at the project level.

Competencies

Spatial planning competencies are nearly everywhere a matter of local competencies. Their implementation and management are also at the same level, even if there are some policies including housing and facilities that are defined at a higher level. Some general guidelines for the organisation of larger areas such as the province, region, county, etc. are in the hands of an intermediate level or of the central government.

Environmental matters are the responsibility of the central government, or shared with the intermediate level. Division of responsibility, sectoral structures and tight compartments are the main obstacles for creating a real integrated system.

Planning, Implementation and Management System

When we are talking about planning we are interpreting things under the lens we built through our knowledge and background. In my own experience, I conceived this trilogy as a wide and interrelated complex structure where every matter related to space and life is considered. This means that every field included has its own approach. Planning has to find the common language and the common bases to integrate and to balance the aims, proposals, priorities, norms, rules, tools, timing, funds, etc. to achieve the goals. I made these two following diagrams to show how I conceive of the functioning system. The concepts surrounded by a gray line are the ones in which environment is included as one more field. The concepts in a gray field are the ones in which the environmentally sustainable principles are mainly reflected (see Figures 5.1 and 5.2).

5.2   Comments

The process being developed in The Netherlands is exemplary. This country has always been a pioneer in the field of organising space. We have learned that '...rule and order is not imposed from above, it pervades the Dutch way of doings things' (Faludi and van der Valk, 1994, pp. 7-8). I will try to stress the points of this process that are of the most significance from my viewpoint.

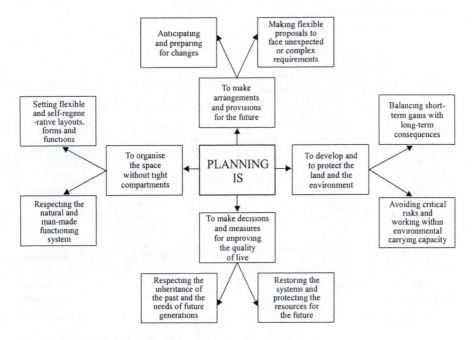

Figure 5.1   Planning system

Main Concepts

It is not good to maintain two types of plans working together. The way to remove the gap between planning and environment is to include both in the same plan, with all the other fields included. This leads to: real planning integration.

It is good to have one central structure defining the major principles for common things as goals, including protection of natural and man-made areas (coasts, natural spaces), standards, etc. But it is important to come to the intermediate level to define more of the generic but geographically specific guidelines. And finally, it is important to provide freedom at the local level for reaching those things which contribute to good quality of life. This leads to a: bottom-up approach within the common general frame.

One important step to overcome the limitations of sectoral decisions and actions is to make projects at the local level. This is a first approach to promote the work in a co-ordinated way. It facilitates the arrangements among the different public and private powers. This leads to: bottom-up co-ordination and integration for planning implementation.

It is important to be able to change direction and to define rules, order and norms with considerable flexibility. We must be prepared to respond to changes quickly. Planning is a process, which means that policy and tools have to be adapted to this characteristic. It is a difficult task but it has to be achieved. This calls for: adaptation and flexibility.

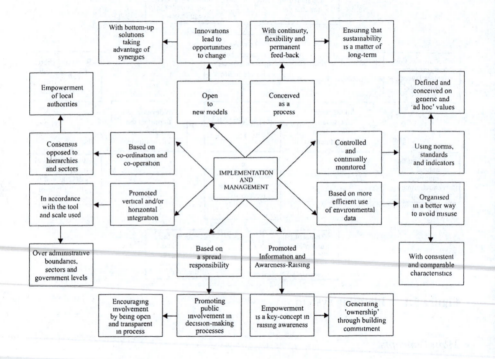

Figure 5.2   Implementation and management system

The ability to control the process at each moment and with continuity is important for the real achievement of the goals in the long-term. This is basic for long-term decision making. Changes, and decrease or improvement in conditions should be quickly identified. This allows the next step to be done in a correct way. This calls for: controlling and monitoring.

To change opinions and lifestyle is another important goal not only for citizens but also for the political and economic structures. The idea of the functioning ecosystem translated into the city system means that development conceived as it was in the past is no longer adequate. The 're' concept – re-use, re-cycled, renewal – requires another pattern. This means that we must be open to: changing patterns.

Indicators and standards are neither general nor unique. Neighbourhoods, cities and states, with their own backgrounds and realities, cannot employ the same indicators. There are some confirmed measures coming from theory and experience, but each of these has to be adapted to the specific situation. This calls for: qualities which are specific for each reality.

Experiences contribute to learning, but solutions have to come from the community engaged in the process. There are no 'best recipes' nor 'the best solution'. Innovations based on exchanged knowledge are especially appropriate. This calls for: experiences, knowledge and synergies.

The exchange of information means collection of data, which is well organised, on a common basis, and easy accessible. This also means the importance of networks of people to facilitate a wider exchange. This point calls for: information and networks.

To decide about goals and the way to achieve them requires knowledge. This implies information and education for all the partners involved in the process. To know and to be aware creates the sense of 'ownership' which allows a deepened and better commitment. This leads to: awareness raising.

Information, commitment and awareness promote participation. This improves the results of the bottom-up approach. If participation is to be comprehensive, the distribution of responsibilities has to change. This means that responsibilities have to be shared among all the partners, public and private, involved in the process. This leads to: participation and spread responsibilities.

The complexity and frailty of the relationship among the different fields require a delicate balance every time we decide to act. We are dealing with a real ecosystem. Any single action will produce reactions, good or bad, in all the fields. Reactions can counteract or complement the desired goal we want to achieve, and destroy or reinforce the unstable balance of the whole structure. This balance cannot be maintained if we act only through sectoral actions. This requires that we understand and deal with: the whole and the parts as a unique entity.

5.3   Conclusion

These twelve points, and the reasoning behind them, are in the view of the author essential to designing an effective strategy for integrating environmental improvement and spatial planning. Current Dutch efforts to achieve this are among the most ambitious of any country in the world, and an important source of guidance for us all.

Note

1   Adriana Dal Cin is Professor at the University Carlos III in Madrid, Spain. She is also Board Member of PLAN & DESIGN, Madrid and of EURONET, Utrecht, The Netherlands. She is Member of the Scientific Secretariat of the Expert Group on the Urban Environment, European Sustainable Cities Project, 1996-98, European Commission, DG XI. Her topics are practice and research on regional and city

planning, natural and cultural heritage protection and spatial planning systems. She is editor (with D. Lyddon) of the International Manual of Planning Practice.

References

Faludi, A., and A. van der Valk (1994) *Rule and Order: Dutch planning doctrine in the twentieth century*, Kluwer Academic Publishers, Dordrecht, The Netherlands.
Miller, D., and G. de Roo (1997) Transitions in Dutch environmental planning: new solutions for integrating spatial and environmental policies, *Environment and Planning B: Planning and Design*, Vol. 24., pp. 427-436.

Chapter 6

Strategies for Protecting Environmental Quality in the Spatial Planning Law of Austria

A. Kanonier[1]

6.1 Introduction

During the 1990s, measures taken to promote environmental conservation have become increasingly important in Austria. Based on a change of values in society which is expressed by a growing sensitivity of large parts of the population towards pollution, conservation regulations have found their way into law on the federal and state levels (Davy, 1996, p. 194; Pernthaler and Fend, 1989, p. 147).

Particularly the increasing ecological damages and the decrease in the standard of living in the cities have brought up new strategies,[2] sometimes leading to restrictive regulations. New regulations have been established for instance for the domains of prevention of air pollution, traffic restriction, water supply, waste treatment, and other major projects (Bachmann et al., 1996, p. 165; Raschauer, 1994, p. 124) aiming to fulfil the requirements of an extensive conservation policy. The increasing demands for open space, especially in and near compact cities, the perceptible limitation of reserve space and the capacity limits of nature led to a strong focus on environmental aspects especially in spatial planning laws and programmes.

One of the main points in connection with improving environmental quality remains an improved control of settlement development. The different utilisation intentions in ground and land which have obviously gained in complexity and caused a lot of conflicts lately, are to be combined in a way that ecological requirements will be taken into consideration for the methodical planning of the settlement structures. It appears that the traditional instruments in the corresponding spatial planning acts of the Länder (states) cannot meet the requirements of a foreseeing and methodical planning of an area anymore. Although the distribution of land utilisation has been regulated legally binding due to the land use plan (Flächenwidmungsplan) in Austria,[3] an uncontrolled urban sprawl has not been prevented or stopped. The consumption of areas for construction purposes remains high. As a reaction to this, the state spatial planning acts have been revised and new regulations have been added in order to do away with the deficiencies that have been recognised. The steps taken to find a solution to the problem are widely differing from state to state.[4] The following will show the reasons for this evolution and the measures that have been proposed in the domain of building activities.

6.2   Open Space and Environmental Quality in Compact Cities

The quality of life and environment in compact cities depends especially on the proportion and the quality of their open space. The various functions of open space in an agglomeration and its direct surroundings can only be briefly mentioned here. In any case it is essential for a city to dispose of enough open space next to heavily built-up areas to be used in compensation. The increasing demand for building areas has led to the misuse of open space for building purposes especially in the immediate vicinity of large cities. In many cases small green land areas which can only fulfil minor ecological, environmental or recreation purposes have been preserved.

The stopping of urban sprawl is one of the main purposes of spatial planning which aims particularly for the best use and protection of the space for living in the interest of public welfare. Corresponding to this task, the state spatial planning acts include more and more principles and aims directed against an increasing urban sprawl (Kanonier, 1996, p. 153). As a conclusion it can be said that existing open space should be used only sparingly for building purposes. The demand for building land is to be realised only on confined areas in order to preserve as large interconnected undeveloped areas as possible.

A very simplified model of regulating building activities can be illustrated as following: Building works are only possible if they correspond to the determinants laid down in the local spatial plans, which are enacted by the municipal council according to the spatial planning acts of the individual Länder (States). As a rule, building works need land appropriately zoned on which the buildings may be constructed.[5] The prohibition of building applies to green land. The decision whether to use a plot as green or building land must be taken by the municipal council. Corresponding to the expected rise in population the need of building land is being determined. It is the task of those responsible for planning to meet the forecast need for building areas in the territory of their municipality (Pernthaler and Fend, 1989, p. 84).

Secondly, the building areas are to be distributed according to spatial planning measures. The owners' building intentions will be limited by public interests and must be in conformity to a predetermined settlement pattern. Finally, high density settlement structures are necessary to grant an economical use of land. From the point of view of spatial planning, the settlement should be as compact as possible.

The realisation of this vision is one of the main tasks of the planning system in Austria. This task not being new, a lot of instruments in various planning regulations already exist and aim for the realisation of a simplified settlement pattern. Most of these instruments are orientated to preserve open spaces and protect them from building measures. This approach is based upon the assumption that if a corresponding protection can be guaranteed, the demand for building land will be performed in unprotected areas. This way of proceeding is based on a very long tradition and has led to a generally matured catalogue of measures to protect open space.

6.3   Protection of Open Space in the Austrian Legal System

The protection of open space is not reduced to one law subject in Austria. Apart from the extensive regulations in the spatial planning acts, other law subjects can be of some importance in particular cases. This is how open spaces can be particularly protected depending on their actual use and need of protection. Interventions causing damage to the environment can be prohibited or prevented for instance by the Forestry Act of 1975[6] or the laws pertaining to water and waterways [7] on a national level or by the nature conservancy laws[8] on the Länder level. These protective regulations apply principally to clearly delimited areas which are particularly sensitive to building interventions or especially at risk because of their location. Woodlands, lakes and rivers and nature reserves in the neighbourhood of large cities are normally under special protection and are not used for settlement activities.

The regulations of the spatial planning acts generally apply to undeveloped properties where no special laws exist for protection regulations. Based upon aims and principles for spatial planning activities, the spatial planning acts provide a hierarchic instrumentarium. These instruments are finally used to decide individually for each property whether it can be built-up or not.

Distribution of Competence in Spatial Planning

In Austria a complicated system of the distribution of powers is laid down in the Federal Constitution of 1929, which divides legislation and administration between the federal government and the Länder (Funk, 1996, p. 113; Walter and Mayer, 1996, p. 114). Municipalities have the right to autonomy, incorporated into the Federal Constitution in 1962 (Pernthaler and Fend, 1989, p. 15). So in Austria spatial planning is a task that is very much split up from the point of view of responsibility.

The present legal situation of spatial planning is based on the distribution of powers as interpreted by the Constitutional Court 50 years ago.[9] The Court held that in principle comprehensive spatial planning is a task within the autonomous sphere of competence of the states, being restricted by the responsibilities of the federation for sectoral planning activities. Consequently the Federal Government and the states and even the municipalities carry out spatial planning activities in parallel. It is important to point out that in Austria the Federal Government does not have the competence for comprehensive spatial planning activities. Comprehensive spatial planning is the responsibility of the states, but it is restricted by sectoral planning measures of the federal government.[10]

Referring to the decision of the Constitutional Court, the states passed spatial planning laws in the 1950s and 1960s. The laws contain goals, contents, procedures, legal effects of the instruments of supra-local as well as local spatial planning activities (Geuder, 1996, p. 23; Hauer, 1984, p. 51).

Aims and Instruments of Spatial Planning

During the 1990s the aims and principles of spatial planning were reviewed by the state legislators. Environmental aspects have gained in importance and are to be taken into consideration at a higher level for the making of decisions (Kanonier, 1994, p. 58). The spatial planning acts differ from state to state but globally the following aims are essential to conservation: space-saving utilisation of land, the stopping of uncontrolled urban spread, better utilisation of reserve building land. It can be concluded that building activities in municipalities should be performed in a limited area and that the other areas should not be built-up. Likewise building in isolated locations within open space should be reduced. These targets are justified in Austria by the lack of space suitable for settlement in many regions. In Tyrol for instance only 13 per cent of the whole territorial area can be used for settlement purposes because of the high percentage of alpine areas. Therefore, the ground is a very limited resource which is to be used carefully in order to provide enough planning liberty for the coming generations (Austrian Conference on Regional Planning, 1992, p. 13).

The practical transposition of these aims occurs on a national, state and municipal level by using particular instruments. As I mentioned before, the federal government does not have the competence for comprehensive spatial planning, but the Federation is responsible for legislation and administration of important sectoral planning activities. Concerning open space, the Forestry Act of 1975 and the Water rights Act are important. For example the respective federal minister is responsible for regulating affairs belonging to forests, and in Austria 46 per cent of the entire territory is covered by forests.

Based on the spatial planning laws, the state government implements spatial planning programmes at the supra-local level. The programmes contain aims of spatial planning in detail and measures necessary for their attainment. Since the end of the 20th century the importance of environmental goals and corresponding measures is increasing. These programmes are set up in form of decrees issued by the state government. The most important instruments on supra-local level are as follows.

State development plans (Landesentwicklungsprogramme) State development plans are drawn up for the entire state territory. Pursuant to spatial planning laws, the comprehensive contents are binding upon the state authorities themselves in implementing regional and sectoral planning, for carrying out their supervisory function over municipalities, and for the spatial planning activities (Fröhler and Oberndorfer, 1986, p. 47). The contents of the state development plans, in fact drawn up in only four states (Austrian Conference on Regional Planning, 1996, p. 137), reach from definitions of goals (for various sectors and planning regions) to measures for developing the structure of the state. State development plans are the level above regional development plans and sectoral plans.

Regional development plans (Regionale Raumordnungsprogramme) Regional development plans are set up for one or more districts or only for parts of districts. They usually contain a mixture of special goals, general guidelines, and corresponding planning measures necessary for further planning work of the municipal planning authorities (e.g. zoning of reservation areas, agricultural zones or centre structures, definition of main tourism centres, maximum settlement limits). The state government is responsible for preparing regional development plans. In some states the municipalities have the possibility to participate in the setting-up procedure. Only a part of the Austrian territory is covered by regional development plans.

Sectoral state plans (Fachplanungen) Sectoral plans are set up for both the entire state territory as well as for individual regions. They deal with spatial planning measures in a limited sector especially for traffic, housing and industry, tourism, zoning of green land, waste disposal, sports facilities, shopping centres. In general, sectoral plans are more detailed than development plans on the state and regional levels due to their limitation to certain sectors and planning tasks. In dealing with open space the Länder have passed various types of sectoral plans with the intention to protect undeveloped areas.[11] In general, they all have the same legal effect: They are binding upon municipalities when zoning building land.

Vienna, the capital of Austria, is both a municipality and a state and for that reason local and supra-local spatial planning cannot be differentiated. Vienna has a special planning system providing for a local development scheme – it is not issued as decree – the land use plan and the building regulation plan are contained in one planning document.[12]

On the municipal level, the most important planning instruments come into being. The local spatial plans are drawn up by the municipalities under their autonomous responsibility (Pernthaler and Fend, 1989, p. 23). Spatial plans on local level are subordinated to supra-local plans and they have to be conform in their contents.

Three types of plans regulated in the spatial planning laws are passed by the municipal council and they can by classified as binding: The local development plan is an internal guideline for the municipal authority, and the zoning plan and the building regulation plan have binding character upon land owners and the municipal authority. In this context the building authority is important, which is responsible for granting building permits.

Local development scheme (Örtliches Entwicklungskonzept) This is a strategic instrument containing guidelines for the development of the whole territory of a municipality and determines the steps necessary for the attainment of these aims on a general level. In some states it has to be produced before a land use plan is laid down (Pernthaler and Fend, 1989, p. 50).

Land use plan (Flächenwidmungsplan) The most important instrument of spatial planning on municipal level is the land use plan. Due to the fact that in Austria each municipality has drawn up a land use plan which contains strong restrictions for land owners it is obvious that the settlement development process is strongly influenced by the local territorial authority (Berka, 1996, p. 70). Land use plans determine in essence the possible use of properties in a municipality. They describe the most rational use of land for the whole territory of the municipality and divide the territory into different utilisation categories, namely building land, green land, traffic area, and special utilisation categories. These main classification categories are further subdivided into various types of utilisation. For example, green land is further subdivided into rural or agricultural areas, recreational areas, camping areas, cemeteries, wasteland, waters and so on. The land use plan must distribute these utilisation types over the municipal area. This distribution has to be based on sufficient and reliable information (Jann and Oberndorfer, 1995, p. 62.). Binding objectives with territorial impact set by the states and the federal government shall be shown in the land use plan (Fröhler and Oberndorfer, 1975, p. 88).

What is the effect of a land use plan? In brief, the plan is binding upon land owners and it influences the possibilities of utilisation for land owners for the future. Conformity with a building land zone in a land use plan is a precondition when applying for a building permit according to the building codes (Mell and Schwimann, 1980, p. 128). If a lot is not zoned as building land, the erection of buildings is not allowed. In principle, construction work within green land is not allowed, but the spatial planning laws contain exceptions for projects which are in connection with agricultural use.

The owner need not change the current use of the lot in case it conflicts with the contents of the land use plan. But if the land owner wants to change the use of the lot it must be in conformity with the plan. The land owner does not have to pay for the zoning of building land, except for the costs of development (Austrian Conference on Regional Planning, 1993b, vol. 105, p. 1).

Building regulation plan (Bebauungsplan)   The building regulation plan is subordinated to the land use plan and determines the use of building land (it is drawn up for building land or parts of it). It contains specific details of construction as limits of buildings, limits of roads and size and density of buildings. In addition to the minimum contents many further stipulations may be included regarding the design of the building. The building regulation plan is also a decree issued by the municipal council and it is binding upon building authorities and property owners. Building regulation plans are drawn up in a scale of 1:2,000 and smaller (Fröhler and Oberndorfer, 1975, p. 105).

In addition to the mentioned instruments, there are instruments of a legally non-binding character, serving as guidelines. Out of this set of instruments the Austrian Regional Planning Concept of 1991 is important. The Austrian Regional Planning Concept was drawn up by the Austrian Conference on Regional Planning and it is addres-sed to the public territorial authorities as bodies responsible for spatial planning. It con-tains a collection of the most important spatial problems in Austria (for example in the fields of settlement policy, environmental policy, regional economic policy), combined with recommendations (Austrian Conference on Regional Planning, 1992, p. 9).

The states and the municipalities are provided with instruments without compulsory function, as well. In individual cases on the state and local levels, spatial planning programmes exist in the form of government resolutions with self-binding effect only upon the administration.

6.4   Settlement Development in Austria

The settlement development especially in the cities has shown an unpleasant evolution lately. Although all municipalities – as mentioned above – dispose of a land use zoning plan, the tendency towards urban sprawl continues. Space is not used carefully. The suburbs are steadily growing and disturbing valuable landscapes. This diagnosis applies to nearly all Austrian cities (Austrian Conference on Regional Planning, 1992, p. 19).

Causes of Urban Sprawl

The causes for the undesirable settlement development are manifold and cannot be completely explained here. However, the causes for the enormous landscape consumption cannot be found in the growth of population. The continuous consumption of areas for residential building is promoted by the increasing requirements for living space (Kumpfmüller, 1989, p. 33). From 1971 to 1991 the expansion of households has been four times higher than that of population. Especially for young and old people the number of two person households is steadily growing. Within 40 years the demand for living space per person has nearly doubled (Austrian Conference on Regional Planning, 1995a, No.121, p. 43). In addition, the detached family house in the country remains the ideal living form and is seen as the real motor of urban sprawl. Further, the trend towards second or holiday homes often endangers sensitive areas in precious landscapes (Schremmer, 1991, p. 45).

Meanwhile the consumption of areas for industrial or commercial purposes has increased. In many cases large businesses are attracted by locations outside existing settlement areas mainly because of lower land prices. Additionally, shopping centres and other large buildings involving a very high volume of traffic are often constructed outside existing settlement areas (Austrian Conference on Regional Planning, 1995a, No.121, p. 43). Generally, these developments have led to settlement patterns which are to be stopped by new instruments in the spatial planning acts.

Deficiencies in the Planning Instrumentarium

The evolution of settlement structures shows very clearly that the measures taken in the spatial planning acts have not been sufficient to separate highly built-up areas from large, interconnected open space. From practical planning experience two main reasons for urban sprawl can be defined. On one hand, the municipalities have consented to important modifications of land use plans in peripheral areas. Based upon the actual wishes of landowners, modifications have been realised which contradict a well-ordered settlement development. On the other hand, buildings have been constructed inside green land because of some special regulations in the spatial planning acts. These buildings do not match with the using of green land. Landowners have for instance pretended to construct a new building in the context of rural activities, but the building were finally used as a second residence (Kanonier, 1994, p. 142).

The state legislators have reacted to this abuse by establishing stricter regulations. The decision of a municipality whether to define ground as building land will now require stricter qualifications. Land use plans (*Flächenwidmungspläne*) can only be modified by municipalities on qualified and provable reasons. These stricter regulations of modifications apply mainly to small isolated areas.

Secondly, in supra-local spatial planning programmes, more and more binding settlement boundaries and green zones or rural priority areas are being defined. These measures have the legal effect that the municipalities are not allowed to zone building land in the marked areas. Finally, special regulations for buildings in green land zones will be drastically reduced and have to undergo a stricter control. The categories of green land are being more exactly defined while building activities are being linked to exact qualifications. For the construction of buildings in green land zones, not only a building permit from the mayor will be required but also an additional permit from the supra-local conservation authority.

From the point of view of the traditional planning system, these measures seem to be successful. Improved protection measures, stricter controls and a reduction of special regulations have always been the typical reactions to an unsatisfactory evolution while realising planning aims. But it has to be noticed that these measures alone will not be sufficient to solve the actual problems.

The stricter measures apply mainly to green land while for a very long time no need for action could have been seen for designated building land. But some of the qualifications required for a successful realisation of these measures belong to the domain of building land. In fact the areas determined as building land are to be employed in building activity. If no building activities are undertaken, the demand for building land will not be reduced which means that the pressure on designated green land remains or even grows. Assuming that the demand for building land is more or less constant, it must be concluded that a successful protection of green land requires that areas defined as building land actually be developed. Strategies aiming only for a limitation of settlement borders will probably fail in the case of an increasing demand for building land.

No Building Obligation

One of the main reasons for the dissatisfaction in matters of settlement development can be found in the system upon which the land use zoning plan is based. While the supply of building land can be regulated by the land use zoning plan, property owners cannot be forced to develop their properties. Even if the zoned building land is fully developed with infrastructure and ready for building, the municipalities has no guarantee that the zoned properties are used in the defined way.[13] It is principally the decision of the owner whether or when building activities will be started on the area defined as building land. Properties in very good locations can persist for many years as green land areas. This situation would be less complicated if the exchange between owners refusing to build and demanders willing to build functioned with the aid of marketing powers and an adequate price (Davy, 1996, p. 197).

Immobility of Building Land

The practical experience shows that a lot of building areas in good locations are vacant but not available.[14] One of the reasons for this situation is the fact that the owners of building areas are not charged in any way, on the contrary, by the classification from green land into building land the property increases in value. From the point of view of the owner the decision not to start any building activities on his property can turn out to be more profitable than any other investment This is even more attractive as the ground prices are steadily increasing.

Individual reasons of landowners for not developing their building land must also be taken into account. Some possible motives might be to take precautions for their own children or some emotional bindings to an undeveloped area. Anyway, the non-conform utilisation of a property leads in many cases to a lack of building sites inside a settlement area. Most of the areas offered for sale are over-priced which means that less wealthy developers are immediately eliminated.

Despite the large amount of land zoned for development by the planning authorities, the demand and the need for building land cannot be fulfilled. The consequence is an increase of the pressure put on the local authorities to zone too much land for building. After some time one part of these areas will be used for building purposes while the largest part will remain free. Too much designated building land, high asking prices for the few sites on the market, and resulting urban sprawl is generally described as the building land paradox (Davy, 1996, p. 197).

6.5   Conclusions: Active Land Policies

This experience is one of the main reasons for the legal reform of the past years. In order to cope with this situation, the new planning law stipulates measures called 'active land policies' to attract building land to the market. The aim of the new regulations is to ensure that zoned building land is used and built up within a certain period of time. But the attempt to find a solution to the problem differs from state to state, primarily depending on the political will of elected officials to add new compulsory measures in the matter of land policy. From the political point of view ground and land are very sensitive domains in Austria (Austrian Conference on Regional Planning, 1993b, No. 105, p. 3; Austrian Conference on Regional Planning, 1995b, No. 123, p. 9; Davy, 1996, p. 195; Kanonier, 1994, p. 181).

In Austria property rights have been constitutionally guaranteed since 1896. The fundamental principle, that 'property is inviolable' is restricted by various regulations in different administrative laws on the federal as well as on the state level, based upon Art. 5 *Staatsgrundgesetz*, which has constitutional status: 'Expropriation against the will of the owner may only occur in those cases and in those procedures as defined by the law'.[15]

Introduction of further compulsory measures in land use planning is a political risk for the decision-makers because the public acceptance of those measures is very low. If a site zoned for building land is not developed within a certain period of time, the new regulations provide sanctions against the landowner. From the point of view of planning law, lots of questions have been raised. It has to be defined for instance when a site has to be developed or what will happen to existing reserves of building land.

As mentioned above, the different Länder have chosen various ways to face the problem. In Styria for example, the obligatory rezoning of building land back into green land is envisaged, in Vienna expropriation is possible. In the Tyrol and Salzburg the importance of property acquisition by municipalities as a measure of 'active land policy' is increasing. Lately municipalities have acquired more and more building land, which is then sold to building co-operatives.

In some Länder, contracts between the land owner and the municipality are envisaged. It will be possible to oblige the owner to enter into a contract under private law with the municipality and to commit the owner to develop the land within a certain period of time. This regulation has been introduced in Salzburg in 1992 and until now good results are reported (Magistrat Salzburg, 1997, p. 3). More plots are available for building purposes while the price of land has been reduced. The spatial planning laws in Lower Austria and Upper Austria stipulate infrastructure fees, which have to be paid every year by the land owners after their property has been zoned for building land.

Which regulation system of measures will be the most successful in matters of acting against the abuse of zoned building land cannot easily be answered. Generally, measures including expropriation, high taxes, automatic rezoning seem to obtain the best results. But they are causing opposition of the landowners and as said before the political realisation can be very difficult. Automatic rezoning of building land within an existing settlement for instance would go against the real planning intentions. The establishment of contracts is only possible with the consent of the landowner. The obligation for all landowners to build up zoned building land would be counter-productive because the precept to use the ground sparingly wouldn't be respected.

Whatever the future evolution of efforts to secure the development of land zoned for building is, since municipalities have the main planning responsibilities, they will have to deal more and more with the questions of land policy and succeed in negotiating the problems of spatial planning more successfully than before. One of the conditions for improved settlement development is a higher sensitivity of the population concerning the need for extensive protection of open space. There is no doubt that stricter measures alone will not lead to satisfactory results. Educational work and supporting measures by the regional administrative body will be required, too.

Notes

1   Dr. A. Kanonier is working at the Department of Law, Faculty of Spatial Planning and Architecture at the University of Vienna, Austria.

2   For a discussion of environmental problems in connection with spatial planning see, e.g., Austrian Conference on Regional Planning, 1988; Austrian Conference on Regional Planning, 1991; Austrian Conference on Regional Planning, 1993; Austrian

Conference on Regional Planning, 1996; Kanonier, 1994, p. 35; Kumpfmüller, 1989, p. 29; Umweltbundesamt, 1988, p. 25; Umweltbundesamt und Österreichisches Statistisches Zentralamt, 1994, p. 179.

3　For a discussion of land use planning in Austria see, e.g., Berka, 1996, p. 69; Fröhler and Binder, 1990, p. 80; Fröhler and Oberndorfer, 1975, p. 88; Fröhler and Oberndorfer, 1986, p. 77; Geuder, 1996, p. 87; Kanonier, 1994, p. 53; Pernthaler and Fend, 1989, p. 44.

4　Spatial Planning Law of Burgenland, LGBl (Legal Gazette, promulgation of acts and decrees) Nr. 1969/18 last amend. LGBl 1994/12; Local Planning Act of Carinthia, LGBl 1995/23; Spatial Planning Law of Lower Austria, LGBl 8000-0 (1977/13) last amend. LGBl 8000-11 (1996/8); Building Code of Lower Austria, LGBl 8200-0 (1996/129); Spatial planning Law of Upper Austria 1994, LGBl 1993/114 last amend. LGBl 1996/78; Spatial Planning Act of Salzburg, LGBl 1992/98 last amend. LGBl 1996/47; Spatial Planning Act of Styria, LGBl 1974/127 last amend. LGBl 1995/59; Spatial Planning Act of Tyrol, LGBl 1993/81 last amend. LGBl 1996/4; Spatial Planning Act of Vorarlberg, LGBl 1996/39; Building Code of Vienna, LGBl 1930/11 last amend. LGBl 1996/44 (W BO).

5　For the building regulation system in Austria see, e.g., Geunder, 1996; Krizizek ,1976; and Mell/Schwimann, 1980.

6　See, e.g., Kalss, 1990, p. 30; Wohanka et al., 1993, p. 29.

7　See, e.g., Raschauer, 1993, p. 25; Rossmann, 1990, p. 19.

8　See, e.g, Bussjaeger, 1995, p. 23; Liehr and Stöberl, 1986, p. 37.

9　VfSlg (collection of decisions by the Constitutional Court) 2764/1954.

10　For a discussion of the decision of the Constitutional Court see, e.g., Fröhler and Binder, 1990, p. 80, Fröhler and Oberndorfer, 1975, p. 59; Geuder, 1996, p. 73; Hauer, 1984, p. 2; Holzer, 1981, p. 49; Jann and Oberndorfer, 1994, p. 58.

11　An interesting example is the regional development program 'Wien Umland', which covers the regions surrounding Vienna. According to the distribution of competencies the regional development program 'Wien-Umland' covers only areas in Lower Austria and does not stipulate planning measures for the City of Vienna. In practice the limitation of responsibility often does not work in a satisfactory way, because the public territorial authorities compete sometimes with one another, intensified by the fact that Lower Austria is ruled by a conservative government, whilst Vienna has a social-democratic mayor. Among others it contains two different types of settlement boundaries. Type one is symbolised with triangles and has the legal effect that municipalities are not allowed to surpass these boundaries when zoning building land. The second type is symbolised with a red hatching and is not an absolute settlement limit. Zoning of building land is only allowed, when building land is reduced in the same amount on another part of the municipality.

12　The first section of the Building Code of Vienna contains regulations for land use and building regulation plans.

13　It is to be noticed in this context that the ownership of land in Austria is distributed over many persons. For a building land area of more than 5000 m^2 not only one but several owners are concerned in most cases.

14　One of the main reasons for urban sprawl and the despoliation of the landscape in Austria can be described with the term 'Building Land Paradox' (Davy, 1996), which contains two elements: on the one hand there are large reserves of unused building land existing and on the other hand building land is not available.

15   For a discussion of land use planning and property rights in Austria see, e.g., Fröhler and Oberndorfer, 1975, p. 153; Fröhler and Binder, 1990, p. 16; Holzer, 1981, p. 216; Korinek and Pauger and Rummel, 1994, p. 161; Pernthaler, 1990, p. 399; Pernthaler and Fend, 1989, p. 39; Walter and Mayer, 1996, p. 499.

References

Austrian Conference on Regional Planning (ÖROK) (1988) *Raumordnung und umfassender Bodenschutz*, ÖROK Enquete 1998, Schriftenreihe No. 64, Vienna.

Austrian Conference on Regional Planning (ÖROK) (1992) *Österreichisches Raumordnungskonzept 1991*, Schriftenreihe No. 96, Vienna.

Austrian Conference on Regional Planning (ÖROK) (1993a) *Siedlungsdruck und Bodenverfügbarkeit*, Schriftenreihe No. 99, Vienna.

Austrian Conference on Regional Planning (ÖROK) (1993b) *Wirksamkeit von Instrumenten zur Steuerung der Siedlungsentwicklung*, Schriftenreihe No. 105, Vienna.

Austrian Conference on Regional Planning (ÖROK) (1995a) *Trends der Siedlungsentwicklung in Österreich*, Schriftenreihe No. 121, Vienna.

Austrian Conference on Regional Planning (ÖROK) (1995b) *Möglichkeiten und Grenzen integrierter Bodenpolitik in Österreich*, Schriftenreihe No. 123, Vienna.

Austrian Conference on Regional Planning (ÖROK) (1996) *Achter Raumordnungsbericht*, Schriftenreihe No. 128, Vienna.

Bachmann, S., K. Giese, Ginzinger, Grussmann, D. Jahnel, M. Kostal, Lebitsch and G. Lienbacher (1996) *Besonderes Verwaltungsrecht*, Springer Verlag, Vienna.

Berka, W. (1996) Flächenwidmungspläne auf dem Prüfstand, *Juristische Blätter*, No. 2, pp. 69-84, Vienna.

Bussjaeger, P. (1995) *Die Naturschutzkompetenz der Länder*, Universitaets-Verlagsbuch-handlung, Vienna.

Davy, B. (1996) Baulandsicherung: Ursache oder Lösung eines raumplanerischen Pradoxons? *Zeitschrift für Verwaltung*, No. 2, pp. 193-208, Vienna.

Fröhler, L, and B. Binder (1990) *Bodenordnung und Planungsrecht*, Institut für Kommunalwissenschaften und Umweltschutz, Vol. 88, Linz.

Fröhler, L., and P. Oberndorfer (1975) *Österreichisches Raumordnungsrecht*, Trauner Verlag, Linz.

Fröhler, L., and, P. Oberndorfer (1986) *Österreichisches Raumordnungsrecht II*, Trauner Verlag, Linz.

Funk, B.C. (1996) *Einführung in das österreichische Verfassungsrecht*, Leykam Kurzlehrbücher, Leykam, Graz.

Geuder, H. (1996) *Österreichisches Öffentliches Baurecht und Raumordnungsrecht*, Linde Verlag, Vienna.

Hauer, W. (1984) *Raumordnungsgesetze der österreichischen Bundesländer*, Prugg Verlag, Eisenstadt, Austria.

Holzer, G. (1981) *Agrar-Raumplanungsrecht*, Österreichischer Agrarverlag, Vienna.

Jann, P., and P. Oberndorfer (1995) *Die Normenkontrolle des Verfassungsgerichtshofes im Bereich der Raumplanung*, Institut für Kommunalwissenschaften und Umweltschutz, Vol. 103. Vienna.

Kalss, S. (1990) *Forstrecht*, Sprinber Verlag, Vienna.

Kanonier, A. (1994) *Grünlandschutz im Planungsrecht*, Orac-Verlag, Vienna.

Kanonier, A. (1996) Baulandwidmung und Zersiedlung im Raumordnungs-recht, *Journal für Rechtspolitik*, No. 3, pp. 151-158, Vienna.

Korinek, K., D. Pauger and P. Rummel (1994) *Handuch des Enteignungsrechts*, Springer Verlag, Vienna.

Krzizek, F. (1976) *System des Österreichischen Baurechts*, Verlag der Österreichischen Staatsdruckerei, Vienna.

Kumpfmüller, M. (1989) *Umweltbericht Landschaft*, Österreichisches Bundesinstitut für Gesundheitswesen, Vienna.

Liehr, W., and Stöberl (1986) *Kommentar zum NOE Naturschutzgesetz*, Orac-Verlag, Vienna.

Magistrat Salzburg – MA12 Bodenpolitik (1997) *Bodenpolitik Salzburg, Die erfolgreiche Lösung eines unlösbaren Problems*, Salzburg.

Mell, W. R., and M. Schwimann (1980) *Grundriß des Baurechts*, Prugg Verlag, Eisenstadt, Austria.

Pernthaler, P. (1990) *Raumordnung und Verfassung (3)*, Universitäts-Verlagsbuchhandlung, Vienna.

Pernthaler, P., and R. Fend (1989) *Kommunales Raumordnungsrecht in Österreich*, Österreichischer Wirtschaftsverlag, Vienna.

Pernthaler, P., and B. Prantl (1994) *Raumordnung in der europäischen Integration*, Literas-Universitätsverlag, Vienna.

Raschauer, B. (1993) *Kommentar zum Wasserrecht*, Springer-Verlag, Vienna.

Raschauer, B. (1994) *Besonderes Verwaltungsrecht*, WUV-Universitätsverlag, Vienna.

Rossmann, H. (1990) *Wasserrecht*, Verlag der Österreichischen Staatsdruckerei, Vienna.

Schmidt-Eichstaedt, G. (1995) *Bauleitplanung und Baugenehmigung in der Europäischen Union*, Deutscher Gemeindeverlag, Köln.

Schremmer, C. (1991) Kein Land in Sicht - Szenarien der Siedlungentwicklung in Österreich, *Der öffentliche Sektor*, No. 4, p. 44.

Umweltbundesamt (1988) *Bodenschutz, Probleme und Ziele*, Bundesministerium für Umwelt, Jugend und Familie, Vienna.

Umweltbundesamt und Österreichisches Statistisches Zentralamt (1994) *Umwelt in Österreich, Daten und Trends 1994*, Umweltbundesamt und Österreichisches Statistisches Zentralamt, Vienna.

Walter, R., and H. Mayer (1996) *Grundriß des österreichischen Bundesverfassungsrechts*, Manz Kurzlehrbücher, Manzsche Verlags- und Universitätsbuchhandlung, Vienna.

Wohanka, E., et al. (1993) *Forstrecht mit Kommentar*, Verlag der Österreichischen Staatsdruckerei, Vienna.

Part B
Regional Approaches to Integration

Chapter 7

Effectiveness of Interjurisdictional Growth Management: Integrated Local, Regional and State Planning in Washington State, USA

R. Piro[1]

7.1   Introduction: Growth Management in an Urban Regional Context

Growth management as a form of urban planning is in a process of evolution in North America. In the 1970s, growth management was a loose grouping of planning programmes and strategies, typically limited in scope and focusing primarily on one issue at time, such as phasing or restricting residential development. The first generation of growth management strategies was also limited in that these programmes and strategies were usually enacted by a single jurisdiction or agency. Over time, growth management has become more comprehensive, addressing more complex planning issues, including traffic congestion, damage to the environment, loss of open space, imbalance between types of land uses, provision of adequate services, and social equity concerns.

Many states, provinces, and urban regions of North America have recognised that growth management is most effective when it is carried out on an interjurisdictional basis, particularly in terms of protecting environment quality. In this chapter, it is argued that a prerequisite to successful growth management is a multi-jurisdictional approach, in which local governments in a single urban region, working together with regional and state agencies, recognise their interdependencies and seek to develop collaborative approaches to managing urban development. A multi-jurisdictional approach is especially challenging in the United States, given the fragmentation of political and decision-making authority within most urban regions. Within a single urban region, there may be hundreds of local units of governments, each with some degree of power or authority affecting land use development and the provision of necessary infrastructure.

This chapter offers a brief overview of growth management as a planning strategy, focusing on the evolution of growth management in the context of regional planning. The primary resources for this portion of the chapter include articles and other literature written on growth management and regional planning efforts. The next section of the chapter is a case study, exploring an integrated growth management planning process developed in the central Puget Sound region. The case study focuses on the policy and plan review process developed by the Puget Sound Regional Council in Washington State and assesses how this process has worked to date. This examination relies primarily on observations by the author, who also has served as project manager of the plan review programme.

The case study examines the benefits and shortcomings of the Regional Council's plan review project, particularly in terms of benefits to local and regional planning programmes, as well as benefits to regional efforts to concentrate development and protect the environment. Problems and deficiencies that have been identified are also discussed, including some that deal with environmental planning issues, such as air quality. The final section of the chapter focuses on lessons learned from the case study, including possible applications to other politically fragmented urban regions. Limitations of the process are described, as well as opportunities for further refining and improving it.

7.2   Growth Management in North America: A Brief Background and History

Since the 1970s, numerous communities across the United States and Canada have sought to develop alternatives to the traditional land use decision-making performed by government and the marketplace, decisions which some community leaders and citizen groups believed to have failed to guide development properly. Simply defined, growth management is a systematic programme of government designed to influence the rate, amount, type, location and/or quality of future development. (For additional definitions of growth management, see Godschalk and Brower, 1989, p. 163; Deakin, 1989, p. 3; DeGrove, 1991, p. xiii; DeGrove and Stroud, 1987, p. 4.)

Beginning in the mid-1980s, many of the single-issue growth management schemes began to be re-examined, with a view toward developing more comprehensive land use strategies and addressing the secondary impacts of growth management policies. Perhaps the biggest problem of growth management during its early development was that interdependencies between land use and infrastructure decisions, particularly across jurisdictional boundaries, were not taken into account.

Within the federated system of the United States, land use decisions – such as zoning – have historically been viewed as belonging to the realm of cities, towns, and counties. Before the 1960s, almost all land use decisions were left to local government, especially municipalities. As a result, only about 25 per cent of the land in the United States was planned for and/or regulated at that time (McDowell, 1986, p. 42).

Emerging environmental issues, as well as concerns over exclusionary actions of local governments to keep out undesirable, yet needed, land uses (such as waste treatment facilities or housing for low income groups), led to what some urban observers called a 'quiet revolution' in land use planning (Bosselman and Callies, 1972). Several states and communities began to view land as a resource having environmental, social, and community value, as well as economic value

(Godschalk, 1986; McDowell, 1986).

There has been growing recognition over the last decades and a half that urban growth and its effects are not isolated to single political units. Land use decisions are not made in a vacuum; there are 'spillover effects' on other jurisdictions. Decisions made by individual jurisdictions frequently create impacts on neighbouring communities and throughout the entire region (Schaenman and Muller, 1975; DeGrove, 1991). Although municipalities are legally independent decision-making entities functioning within their distinct political boundaries (Perin, 1977), local communities are also faced with regional issues, including economic development, the efficient provision of services, social equity, and environmental quality. In many urban areas of the United States and Canada, regional problems, such as traffic congestion or waste water treatment, have become so acute that the issue of local control and decision-making within urban regions must be confronted.

Intergovernmental Growth Management

Interjurisdictional programmes and strategies for growth management are emerging in a number of urban regions across North America, although they vary significantly. Some regional programmes are voluntary efforts, developed around a narrow set of common planning goals or a few mutually agreed-upon issues, such as the provision of adequate services or co-ordination of open space planning. Other programmes are more structured planning efforts, frequently responding to state-mandated legislation for integrated planning programmes, in which regional and local growth management programmes must demonstrate consistency with state planning regulations or goals.

The mechanisms for considering regional interests have been somewhat fluid, particularly in urban regions of the United States. Nevertheless, there is a growing acknowledgement that growth-related problems cannot always be adequately handled by a single jurisdiction. In several metropolitan areas across the country, planners, decision-makers, and business leaders have been examining ways to deal with urban growth and environmental issues in a more comprehensive and regional manner.

Several of the more comprehensive regional strategies share three key objectives: (1) interjurisdictional co-ordination; (2) concurrency – that is, ensuring needed facilities are in place at the time development occurs; and (3) compactness – that is, concentrating development in more compact communities (DeGrove, 1992).[2] Many urban observers agree that the key to successful growth management is providing consistency in interdependent planning efforts, both vertically among various levels of government and horizontally across neighbouring jurisdictions.[3]

7.3 The Case Study: The Central Puget Sound Region of Washington State, USA – an Integrated Model for Regional Growth Management

One type of programme, implemented in the central Puget Sound region of Washington State, is an integrated approach to growth management. Local governments operating under revised state law (i.e., the Washington State Growth

Management Act, adopted in 1990, amended in 1991 and 1994), are now required to develop comprehensive growth management strategies that address the protection of resource lands and critical areas, preservation of rural lands, compact and concentrated development, transportation, capital facilities and utilities planning, and economic development. These plans are to address broad state planning goals and must demonstrate consistency with other local and regional planning efforts. Interjurisdictional consistency in growth management planning is now a requirement, although how this consistency is actually accomplished was left to local and regional officials to define and develop.

Consistency Requirement for Local and Regional Planning

Prior to 1990, comprehensive planning was required only in Washington State if a local jurisdiction wanted to implement zoning. The only elements mandated to be in a comprehensive plan were (1) a land use element, and (2) a 'circulation' element to address transportation. There was little direction provided by the state as to what was to be addressed in these elements, and no requirement for consistency (WSDOT, 1989).

By the late 1980s, rapid growth and increasing traffic congestion provided an impetus for requiring a more rigorous and comprehensive approach to planning and the management of growth.[4] During the 1980s, land was being developed in the Seattle area at over twice the rate of actual population growth.[5]

In 1985, King County – which includes the City of Seattle and now more than 30 municipalities – adopted a comprehensive plan for its unincorporated areas. This plan included a number of growth management provisions (King County, 1985).[6] For growth issues requiring intergovernmental co-ordination, the county committed to establishing interlocal agreements with its municipalities. These agreements were to provide for a common commitment and co-ordinated approach to working together on major planning and growth management concerns. The process for developing these agreements became laborious and often resulted in adopting only very modest commitments to co-operate. There was virtually no oversight authority to make sure the parties to these agreements would actually uphold their commitments. This experience led the county, other local jurisdictions, and state officials to work with the Washington Legislature to explore developing a more effective way to co-ordinate local planning efforts to meet regional growth management concerns.

With the adoption of the Washington State Growth Management Act in 1990, consistency is now required where there are 'common borders or related regional issues' (Revised Code of Washington, Chapter 36.70A.100). The Act identifies three areas of consistency: (1) internal consistency, which addresses the consistency of various elements within a comprehensive plan with one another; (2) consistency of regulations and zoning with the adopted comprehensive plan and its policies; and (3) intergovernmental consistency among the plans adopted by various jurisdictions and planning agencies. The focus in this study is on this third form – that is, intergovernmental consistency.

Once consistency became an actual requirement, a multitude of co-ordination task forces, technical committees, and advisory committees were formed in the central Puget Sound area. By 1992, Regional Council staff had identified more than 100 such co-ordination groups operating in the four-county region. These groups often were charged with dealing with a single planning issue that required intergovernmental co-ordination, such as establishing level-of-service standards, designating urban growth areas, developing compatible information and databases, or addressing affordable housing needs. There was plenty of co-ordination going on, but there was no overall co-ordination of all the various co-ordination groups.

In response to demand for more structured co-ordination, the state legislature amended the Growth Management Act in 1991 to require development of countywide and multicounty planning policies. These multi-jurisdictional policies serve as the planning framework for all municipal and county comprehensive plans within the urban region.[7] These policies are to address: (1) urban growth areas, (2) contiguous and orderly development, (3) regional capital facilities, (4) interjurisdictional planning, (5) transportation, (6) provision of affordable housing, and (7) economic development. Local comprehensive plans are required to reflect the adopted countywide and multicounty policies (RCW, Chapter 36.70A.210). Countywide policies were adopted within all the counties of the central Puget Sound region in 1992 and some have been recently amended. The multicounty policies for the entire region were adopted by the Puget Sound Regional Council's General Assembly in 1993 and updated in 1995.

Puget Sound Regional Council: Mission and Authority

The Puget Sound Regional Council serves as the regional growth management, economic, and transportation planning agency for the four-county central Puget Sound region in Washington State, which includes the central cities of Seattle, Tacoma, and Everett. The four counties are King, Kitsap, Pierce and Snohomish. There are 80 cities within the four counties, with a combined population of more than three million. The region covers 2,500 hectares (6,400 square miles) and is quite diverse, including urban, rural, suburban and resource lands. The region abuts Puget Sound, an inlet of the Pacific Ocean and includes several natural forests in the Cascade Mountains. Mount Rainier, located in a United States national park, is also within the region and is more than 4,500 metres (14,000 feet) in elevation.

The Puget Sound Regional Council is one of several Regional Transportation Planning Organisations (RTPOs) designated in Washington State according the Growth Management Act. The Regional Council also serves as the Metropolitan Planning Organisation (MPO) for the four-county region, one of hundreds of such organisations established under United States federal law. These planning organisations are res-ponsible for transportation planning and the distribution of federal transportation funds to support locally-sponsored and regional transportation projects. The Puget Sound Regional Council, formed in 1991, is a membership organisation with an Executive Board of elected officials from local jurisdictions within the region. The Council is responsible for developing and maintaining the regional growth strategy, VISION 2020, and its transportation component, the Metropolitan Transportation Plan (MTP).

VISION 2020, updated in 1995, is the long-range growth management, economic and transportation strategy for the four-county central Puget Sound region. The Multicounty Planning Policies, required under the Washington State Growth Management Act, are incorporated in VISION 2020 and provide the policy framework for local, regional, and state planning efforts in the Puget Sound region. VISION 2020 calls for containing most of the region's growth within urban growth areas, creating compact urban communities, and focusing growth into urban centres and in manufacturing/industrial areas. The strategy intends to conserve farmlands and other natural resources, as well as to provide necessary services and facilities in an efficient and cost-effective manner.

The Metropolitan Transportation Plan (MTP), also adopted in 1995, is VISION 2020's more explicit transportation component. This plan outlines a long-term approach for dealing with the mobility of people and goods throughout the urban region. The Metropolitan Transportation Plan reinforces the VISION 2020 growth strategy by emphasising transportation programmes and improvements that are multimodal and increase travel options.

Regional Consistency Review

To help ensure that local growth management efforts demonstrate consistency with regionally adopted plans, regional planning agencies in Washington State have been given the authority to review local plans for conformity with the adopted regional plan. The focus of this regional review is on the consistency of local comprehensive plans, transit agency plans, and the state transportation plan with regional transportation policies and provisions. The review process recognises the pivotal role of the regional plan in the development of land and the provision of adequate services.

The consistency review process being implemented in the Seattle-Tacoma-Everett urban region by the Puget Sound Regional Council respects the tradition of local authority and decision-making so important in the North American context, while challenging local governments to acknowledge their regional responsibilities for managing growth and protecting the environment. To make the process more meaningful, only jurisdictions whose plans have been certified are eligible to participate in the regional distribution of federal dollars for transportation improvements projects.

Developing the Consistency and Co-ordination Review Process

The Regional Council's plan review project focuses on assessing how well local, countywide, transit agency, and state transportation planning efforts address regional planning concerns. The goal of consistency is not to force uniformity among jurisdictions, but to reduce conflicts and make sure various planning efforts are compatible. To assure consistency in transportation, the Washington State Growth Management Act authorises the Regional Council to certify transportation elements for conformity with state transportation planning legislation and consistency with the adopted regional transportation strategy.

Regional Council staff worked with local planning staff for six months to develop a plan review process that would allow regional review of, and comment on, local plans. The process, adopted in mid-1994, has three distinct parts: (1) certification review of transportation elements in local comprehensive plans for conformity with the state Growth Management Act requirements for transportation; (2) consistency review of multicounty policies and the countywide planning policies; and (3) co-ordination review of multicounty policies and local comprehensive plans.

Certification review is the real muscle in the review process, providing a detailed analysis of the transportation provisions in a local jurisdiction's comprehensive plan. The transportation provisions in the countywide planning policies are also subject to certification. The local transportation element and related provisions are reviewed to determine if they meet the various planning requirements in the Washington State Growth Management Act, including consistency with the transportation provisions in the region's VISION 2020 strategy. The countywide planning policies are reviewed primarily to determine if their provisions are consistent with the Metropolitan Transportation Plan.

In addition, the certification process has been tied to the Regional Council's responsibility for distributing federal transportation dollars to jurisdictions and agencies seeking funding for regional transportation projects and improvements. Jurisdictions whose plans were not certified by 1997 were not eligible to compete for transportation dollars during the 1998-2000 regional Transportation Improvement Programme (TIP) administered by the Regional Council.

Consistency review of multicounty and countywide planning policies is designed to provide analysis of the compatibility of the various framework policies in place to guide the development of local plans. Co-ordination review is a more informal review of other growth management and economic development issues that might be addressed in local, transit agency, and state plans. As part of co-ordination review, provisions in local, state, and other regional plans are compared with VISION 2020 policies on urban growth, housing, service provision (including transit), and economic development. Comments and recommendations are provided back to jurisdictions and agencies to guide their future planning efforts.

Procedures

The plan review process follows a simple schedule. (1) Once a jurisdiction, transit agency, countywide organisation, or the state Department of Transportation adopts policies or a plan, the completed planning document is submitted to the Regional Council. (2) A questionnaire on certification and co-ordination issues developed by Council staff is sent to the jurisdiction or agency. This questionnaire is designed as a self-evaluation checklist for working through various requirements and policy issues. (3) Regional Council staff review the information provided in the questionnaire and prepare a draft certification and consistency report. The report summarises how well jurisdictions, countywide organisations, and agencies have addressed the various criteria outlined in the questionnaire. The draft report is sent to the jurisdiction or agency for staff review and comment. (4) The report is then revised and transmitted to the Regional Council's Transportation Policy and Executive Boards for consideration. When a local comprehensive plan or set of countywide planning policies is reviewed, a recommendation for certification of the transportation provisions is also incorporated into the report. (5) The Executive Board completes the review process when it acts on the recommendation to certify the transportation element in the local comprehensive plan or the transportation provisions in the countywide planning policies.

Transit agency plans and the state transportation plan are not currently required to be certified. For those plans, comments and recommendations are provided back to the planning agencies to guide their future planning activities.

Certification Review: Conforming with the requirements of the Growth Management Act

The Regional Council uses five criteria in reviewing transportation elements in local comprehensive plans for conformity with the transportation requirements in the Growth Management Act. A brief description of these criteria follows.

Land use assumptions used to develop the required traffic volume estimates and transportation modelling Regional Council staff looks for linkages between population, housing, and employment forecasts and the traffic and transportation estimates presented in the comprehensive plan.

Inventory of transportation facilities and services The transportation element is reviewed to determine whether the jurisdiction has identified key transportation facilities, both current and future. In addition, the element is analysed to see if level-of-service standards have been established for arterial roadways and transit service.

Funding strategies to pay for needed transportation facilities and services The funding component must identify both cost estimates and probable sources for financing transportation improvements. The transportation element is also required to include a reassessment strategy to address situations in which the jurisdiction might find that it has a shortfall in funds needed to finance transportation facilities and services.

Interjurisdictional co-ordination The transportation element is reviewed to determine if the jurisdiction has linked its transportation planning efforts with those of adjacent jurisdictions. In particular, the plan is examined to see if the jurisdiction took into account the land use and transportation impacts of its transportation system on neighbouring cities and counties and vice versa.

Transportation demand management strategies The Council reviews the transportation element, looking for programmes and strategies designed to reduce the rate of growth in the number of automobile trips, especially single-occupancy vehicle trips. (Transportation demand management is also a key policy area in the Metropolitan Transportation Plan.)

Certification Review: Consistency with VISION 2020's Transportation Provisions, the Metropolitan Transportation Plan

In assessing whether the transportation elements and related provisions in local comprehensive plans are consistent with VISION 2020, the Regional Council focuses on the four major transportation policy areas in the Metropolitan Transportation Plan, which is the detailed transportation component of VISION 2020. The Council also reviews plans for conformity with state air quality requirements. (Portions of the central Puget Sound region are currently designated as a 'maintenance region' for air quality under United States federal provisions.) These four policy areas and the related air quality provisions also are used in the review of countywide planning policies, transit agency plans, and the state transportation plan. (These policy areas are the primary focus of the certification review performed on the countywide planning policies.) These policy areas are detailed below.

Maintaining and preserving existing transportation facilities Local, regional, and state policies are reviewed for financing strategies and routine maintenance programmes designed to preserved various transportation facilities, such as roadways, transit facilities, and bikeways. Transportation system management programmes (such as signalisation improvements, access and turn restrictions, and other capacity enhancement strategies) are all expected to be included in local transportation elements.

Transportation demand management Policies and plans are reviewed to determine if they include programmes and strategies designed to reduce automobile usage, particularly single-occupant vehicles. Such strategies may seek to improve transit use, as well as walking and bicycle access to transit facilities, employment areas, and other destinations. Other programmes may focus on reducing commuter trips, including ridesharing programmes and the promotion of telecommuting.

Co-ordinating land use and transportation planning VISION 2020 stresses that the way we use and develop land significantly impacts our overall mobility. Policies and plans are reviewed to see if they include policies and provisions to concentrate development in compact areas and provide opportunities to develop housing near employment areas and retail locations. In addition, policies and plans are reviewed to determine if jurisdictions and agencies have addressed urban design in a manner that supports transit use and walking, such as siting buildings to allow easy pedestrian access to transit stops.

Expanding transportation capacity. The Metropolitan Transportation Plan's focus on capacity expansion emphasises improving connections between different modes of travel. An example might be the inclusion of projects or provisions designed to complete connections within the region's non-motorised transportation network. Policies and plans are also reviewed to determine whether they have addressed the movement of freight and goods.

Air quality conformity The review process is used to look for both policies and provisions committing the jurisdiction or agency to conforming with state air quality requirements. The air quality provisions should expressly address reducing criteria pollutants. Examples of such provisions include trip reduction strategies and ordinances, employer-based transportation management programmes, flexible work schedules, ridesharing programmes, parking management, improved transit and non-motorised facilities, and restrictions on vehicle use.

Consistency and Co-ordination Review – Compatibility of Plans and Policies with other Issues Addressed in the VISION 2020 Strategy

The more informal consistency review of countywide policies and co-ordination review of local comprehensive plans primarily addresses the other major non-transportation policy areas of the VISION 2020. These policy areas include (1) planning for needed capital facilities, services, and utilities, including the siting of needed regional facilities; (2) providing adequate housing for all income groups and for people with special needs; (3) economic development; (4) protecting resource lands, critical areas and open space; and (5) establishing urban growth areas. Comments, suggestions, and recommendations on these topics are included in the consistency report that is transmitted to the local jurisdiction, although these particular comments are not factored into the recommendation for certification of the transportation elements. If any inconsistencies are identified between countywide or local planning policies and the VISION 2020 strategy, the Regional Council then engages in a reconciliation process with the countywide organisation or local jurisdictions to develop more compatible policies and/or provisions.

Initial Results

In the first two years of reviewing policies and plans, the Regional Council prepared consistency reports for four sets of countywide planning policies, 76 municipal and county comprehensive plans, and three transit agency plans.[8] The countywide policies for three of the counties were certified for consistency with VISION 2020. Action was delayed on certifying the policies for the fourth county, Kitsap County, primarily because its policy provisions were different in character from the multicounty policies in VISION 2020 and from the other three counties' policies. Kitsap County is examining ways to revise its countywide policies to better reflect the type of policies adopted by the other counties in the region and by the Puget Sound Regional Council.

When initially submitted, the Pierce County's Countywide Planning Policies were 'conditionally certified' by the Regional Council. The county's policies failed to address one of the four major policy areas in the transportation provisions in VISION 2020, namely the provision to encourage more concentrated development as a strategy to improve mobility and accessibility. Pierce County subsequently amended its policies to incorporate policies and provisions calling for the development of urban centres and areas of concentrated development that support transit use, bicycling, and walking. Once these amendments were adopted, the Countywide Planning Policies were fully certified by the Regional Council.

Approximately two thirds of the transportation elements in municipal and county comprehensive plans were certified following their adoption and initial review by the Regional Council. These elements adequately demonstrated conformity with the Growth Management Act and consistency with the Metropolitan Transportation Plan.

During the initial round of review, the Council's Executive Board conditionally certified the transportation elements in four municipal comprehensive plans. Conditional certification recognises that a transportation element has satisfied most of the Growth Management Act requirements and may only need modest additional work in one or two policy areas in order to be in full compliance with the Act.

Twenty other municipalities were offered conditional certification or the option to correct deficiencies in their plans and then resubmit them to the Regional Council for consideration. For three localities, the Regional Council has recommended delaying action on certifying their transportation elements. These municipalities' plans have a number of deficiencies which need to be corrected before the transportation elements can be recommended for certification. All jurisdictions were required to bring their plans into full conformity by the 1997 deadline to be eligible to participate in the 1998-2000 regional transportation improvement programme.

In reviewing the transit plans for three of the transit agencies serving the central Puget Sound area, the Regional Council found all three plans to be consistent with the Metropolitan Transportation Plan. Included in this review was the long-range plan for the newly created Regional Transit Authority (RTA), charged with developing a high-capacity transit system for King, Pierce and Snohomish counties.

7.4   Lessons, Limitations and Opportunities

This portion of the chapter is an evaluation of lessons learned from the case study. Also included is a discussion of the limitations of the Puget Sound programme, as well as suggestions for improving the process and possible applications to other politically fragmented urban regions.

Problems and Deficiencies

Most of the comprehensive plans reviewed have done an excellent job of incorporating VISION 2020 policies and principles into their various plan elements. Where problems or deficiencies have been identified, they relate more to inadequately satisfying state planning requirements outlined in the state Growth Management Act.

Four significant problem areas have emerged in the initial review of policies and plans. The first area involves the diverse approaches to establishing level-of-service standards for arterial streets. Although most local jurisdictions have satisfied the requirement to establish level-of-service standards, initial analysis indicates that cities and counties in the central Puget Sound region have used a wide variety of approaches in determining standards. For example, some jurisdictions are using traditional volume-to-capacity ratios, while other jurisdictions have developed innovative standards that factor in land use assumptions and travel time. There is a concern that the different methodologies employed by individual jurisdictions may not always be compatible from a regional perspective.

A second problem area for jurisdictions is the Growth Management Act requirement to establish level-of-service standards for transit service. The problem here is that jurisdictions themselves do not provide transit service; these standards are typically set by transit agencies that provide service to an entire county or to a large number of jurisdictions within a county. In this situation the Regional Council has recommended that jurisdictions simply reference the service standards or guidelines already established by their respective transit providers.

A third problem commonly found in the review of local plans is the failure of some jurisdictions to link their transportation finance strategy with the concurrency requirement spelled out in the Growth Management Act. The concurrency requirement mandates that jurisdictions ensure that adequate infrastructure, or a financial commitment to fund needed improvements, be in place at the time development occurs. As a result, jurisdictions are required to address how they might handle potential shortfalls in funding for needed transportation improvements. The Regional Council encourages jurisdictions that have not established strategies for potential funding shortfalls to consider three options: (1) identify potential sources for additional funding dollars; (2) adjust their level-of-service standards to a lower level; and/or (3) revise their land use strategies to phase development to occur at the time facilities and infrastructure can be built.

Finally, a large number of jurisdictions have not adequately addressed federal and state air quality conformity requirements. This problem is partially due to the fact that the state requirement to address air quality in local growth management plans is not mentioned in the state Growth Management Act, but in the state Air Quality Conformity legislation. In situations where this requirement has not been addressed, the Regional Council has recommended conditional certification of the local jurisdiction's transportation element, and has asked the affected municipalities to add the required air quality policies and provisions by the 1997 deadline for certification.

Benefits to Local and Regional Planning Efforts

Besides meeting Washington State requirements to have a certified transportation element, certification also allows jurisdictions to submit transportation projects for funding through the Regional Council's transportation improvement programme. This programme is a competitive process for federal funds allocated to the region for transportation improvements which are deemed to best satisfy transportation and policy criteria established by the Regional Council.

Through the plan review process, cities, counties, and transit agencies can also gain regional support for their local planning efforts. This endorsement can be very valuable for jurisdictions, as well as countywide organisations and other agencies, as they seek to develop a support base for their plans and policies. This support base includes citizens and various interest groups, elected officials, planning commissioners, and other decision-makers, such as state transportation officials.

Policies and plans that are found not to conform with the Washington State Growth Management Act, or are determined to be inconsistent with VISION 2020 and the Metropolitan Transportation Plan, are potentially subject to state sanctions that would curb some of the jurisdiction's local taxing authority.

Previous regional development or transportation plans were often forgotten or ignored by local jurisdictions in preparing their own comprehensive plans. Now regional policies and provisions are routinely addressed in local plans. VISION 2020 and the Metropolitan Transportation Plan are given serious attention at the local level.

In addition, local plans are now taken into account when considering further refinements of regional policies and plans. Information obtained in the review of local and agency plans and policies will be used in developing the next update to VISION 2020 and the Metropolitan Transportation Plan. The review process provides an opportunity to identify what local jurisdictions are doing that ought to be incorporated in the regional plan. Local plans and the regional plan have become mutually reinforcing.

Limitations of the Process

The consistency review is not used to review development regulations and zoning. This may be a significant limitation, since zoning and regulations are the primary tools used to implement local comprehensive plans. However, the Regional Council is developing another project, performance monitoring, which is designed to evaluate how the region overall is faring in achieving the key provisions in the VISION 2020 strategy. The monitoring project is being developed in conjunction with similar performance evaluation projects initiated by several of the cities and counties in the region.

By subjecting only the transportation element and related provisions to certification, the process does not always reveal the shortcomings in other portions of the plan. Individuals who support a more narrow review process argue that since all elements of a plan are to be mutually reinforcing (that is, internally consistent), then certification of just one element is sufficient. After all, the transportation element is required to build on the land use assumptions in the comprehensive plan, including population and employment estimates, as well as the preferred land use concepts. Under a 1994 amendment to the Growth Management Act, transportation elements already reach beyond traffic issues, since they are now required to address housing densities, mixed-use development, and economic development.

However, there may be a problem with this argument when it comes to environmental planning issues, other than air quality. In particular, the importance of protecting critical areas is not discussed in any detail in the Act's requirements for transportation elements. The Regional Council frequently recommends that local jurisdictions better integrate these requirements if they have not done so in their adopted plans; however, the lack of such provisions does not preclude certification.[9]

The 1994 amendment to the Growth Management Act has now established a Growth Management Hearings Board for the central Puget Sound area, which has the authority to hear charges against jurisdictions for failing to conform with the various planning requirements, including protecting and preserving resource lands and critical areas. A number of plans have already been remanded back to jurisdictions by the hearings board for inadequately addressing population forecasting requirements or land use requirements, including those that are intended to address environmental concerns.

Opportunities for Further Refinement and Improvement of the Process

Some local staff and officials expressed strong concern when the Regional Council first began reviewing plans, particularly since the first few certification actions included a recommendation of conditional certification, as well as a recommendation to delay action on one county's countywide planning policies. The Regional Council responded by agreeing to conduct an evaluation of the policy and plan review project one year after it began. As part of this evaluation, the Council conducted a workshop with interested staff from jurisdictions.

The workshop participants agreed that they liked the thoroughness of the Regional Council's Consistency and Certification Reports. Several staff indicated that the comments in these reports had been incorporated into the jurisdiction's work programme and were definitely influencing local planning efforts and decisions.

Workshop participants also were asked to share suggestions for improving the policy and plan review process. Some local staff encouraged the Regional Council to provide comments early in the review of amendments and updates to policies and plans. It was agreed that it would be easier to modify or change amendments while they are still in draft form, even though the Regional Council's responsibilities are clearly directed at reviewing adopted policies and amendments. A number of local staff recommended that the Regional Council should strengthen the comments and recommendations it makes when discussing problems identified during the review process. The Council was encouraged to provide more direction and guidance on suggesting steps jurisdictions and agencies should take to improve their policies and plans.

Some county staff recommended that the Regional Council provide more assistance to smaller jurisdictions to help them through the plan review process. These smaller municipalities do not always have the staff resources to work through all of the requirements outlined in state law.

Next Steps

At the workshop, the Regional Council was also encouraged to develop guideline reports on topics frequently identified as problems or deficiencies in the consistency review process. In these reports, Regional Council staff highlight key legislative requirements, provide guidance on policy options for local government to consider, and cite examples of good policies adopted in plans that have already been certified. One such guideline report was completed for air quality conformity and two others are currently being developed to address financing requirements and the application of land use assumptions in traffic volume estimation and transportation modelling.

The Regional Council is also preparing composite reports on how various planning issues are being treated in the region. For example, how is the region as a whole faring in implementing key legislative mandates, such as developing transportation demand management strategies? These reports will be of value to various planning agencies and organisations as work proceeds on refining and updating local comprehensive plans, transit agency plans, countywide planning policies, and the state transportation plan. They also will help to enhance other projects and programmes being undertaken by the Regional Council, including its update of VISION 2020 and the Metropolitan Transportation Plan; the performance monitoring project, designed to track the implementation and effectiveness of key regional policies and strategies; and the regional transportation improvement programme, which is partially based on how well transportation projects submitted for federal funding reinforce various regional policies and programmes.

As the Regional Council continues its review of amendments and updates to local plans and countywide policies, it is anticipated that the review process will become stronger and more detailed with each subsequent round of review. Finally, the Regional Council plans to set up a 'clearinghouse' to track the most effective programmes and strategies used in planning efforts throughout the region. This clearinghouse function will help to identify some of the 'best practices' and provide technical assistance back to jurisdictions and other planning organisations and agencies to help advance their individual and collective planning efforts.

The Regional Council's Growth Management Programme as a Model for Other Urban Regions

With cautious optimism, the Puget Sound Regional Council's policy and plan review project may well become a model for integrating local and regional planning useful for other urban regions working with a growth management framework for planning. The Puget Sound region has found a way to make regional planning meaningful at the local level, where the primary implementation of projects and programmes continues to occur. The Council's policy and plan review project builds on the recognition that local, regional and state planning efforts are interdependent and must be pursued co-operatively in order to be successful. The Regional Council's process also has been able to better integrate local planning into the development of regional planning policies and strategies. Indeed, this process may well have pioneered the way for building more solid intergovernmental relationships for future planning efforts undertaken by various governmental agencies and planning organisations. The sensitivity and delicate diplomacy built into the Regional Council's review process, along with its linkage of certification to access to federal transportation dollars, may serve well as a model for balanced consistency review in other politically fragmented urban regions, even those not planning under a state mandate.

7.5   Conclusion

The Puget Sound Regional Council's process has already demonstrated some success in addressing some of the key objectives of regional growth management, especially in the area of intergovernmental consistency. The Regional Council's plan review process has improved co-ordination between local, county, regional, and state planning efforts, especially in the area of transportation planning. The review process is used to determine whether jurisdictions have met the concurrency requirements established in the Washington State Growth Management Act. Local comprehensive plans must demonstrate to the Regional Council that they have linked their land use and transportation planning elements. Jurisdictions must show that they can provide the infrastructure needed to support their land use strategies and development plan.

The review process has had more limited results with non-transportation issues. Although local jurisdictions are required under state law to identify resource land and critical areas and to develop regulations to oversee development in these areas, there are not clear connections in the law guiding how these requirements are related to the individual elements in comprehensive plans. The Regional Council's review process is itself limited, with its primary environmental focus being on air quality. The overall planning process, from the local level to regional and state agencies, would be greatly enhanced if the certification requirement was extended to apply to the review of all required elements in local plans.

The consistency requirement in the Growth Management Act has led to the development of a continuous process in which there is ongoing interaction between local and regional planning efforts. The Regional Council's review process has advanced the co-ordination of local, regional and state planning efforts to a level beyond any previous attempts to better integrate land use and transportation planning in the central Puget Sound region. Indeed, the Council's review process has become a mechanism for ensuring consistency, while continuing to respect local control.

Notes

1   Rocky Piro is a senior growth management planner at the Puget Sound Regional Council in Seattle, Washington, USA. He manages the Regional Council's Policy and Plan Review Project and was project lead on the 1993 Update to VISION 2020. Previously, he was project manager for Intergovernmental Planning Team for the King County Department of Parks, Planning and Resources, also in Washington State. Dr. Piro has also taught courses in intergovernmental planning and community planning at the University of Washington in Seattle.

2   For a more detailed discussion of various state-mandated growth management programmes, see Bollens, 1992; Brower, et al., 1989; DeGrove, 1992; Gale, 1992.

3   Examples of urban regions in the United States exploring integrated growth management strategies: Denver (Colorado), Minneapolis-St. Paul (Minnesota), Portland (Oregon), Sacramento (California), San Diego (California), San Francisco (California), and Seattle Washington), as well as the states of Florida and New Jersey. For further reading, see Bay Area Council, 1988 (San Francisco); Bollens, 1992 (Florida); Denver Regional Council of Governments, 1996; DeGrove, 1992 (Florida, New Jersey, Portland); Detwiler, 1992 (San Diego); DiMento and Grayer, 1991 (Minneapolis-St. Paul); Epling, 1992 (New Jersey); Gale, 1992 (Florida, New Jersey)); Johnson, et al., 1984 (Sacramento); Keefe, 1992; Metropolitan Council in the Twin Cities Area, 1986 (Minneapolis-St Paul); Piro, 1993; Rapaport, 1992a (San Francisco); Rapaport, 1992b (Minneapolis-St. Paul); San Diego Association of Governments, 1992. Canadian examples include Toronto (Ontario) and Vancouver (British Columbia).

4   Traffic studies conducted by the Texas Transportation Institute indicate that the Seattle-Tacoma-Everett urban region is the sixth most congested urban region in the United States. Severe auto congestion contributes to environmental and health problems, as well impacting the region's overall quality of life.

5   According to research published by 1000 Friends of Oregon, population in the Seattle area grew by 38 per cent between 1970 and 1990, while land area grew by 87 per cent. For a more detailed discussion of expansion in land area and population in Seattle and other selected metropolitan areas in the United States, see Kasowski, 1992.

6   Since the 1930s there have been a number of efforts to create a more regional approach to governance in metropolitan Seattle. There were proposals to consolidate county and city governments into a single political entity (1933), allow Seattle as the most urbanised portion of the region to separate from King County and form a distinct urban county government (1945), transfer urban planning and governance authority to the county (1956), form a separate municipal planning organisation to oversee co-ordinated planning for Seattle and its growing suburbs (1958), and create a voluntary association of local governments to co-ordinate regional planning issues (1958, 1967, 1974) (Washington State Local Governance Study Commission, 1988).

7   At the end of the 1990s, the central Puget Sound region is the only urban area of the state required to develop both countywide and multicounty planning policies.

8   As of March 1997, 45 transportation elements in local comprehensive plans had been certified by the Regional Council's Executive Board. Six more were scheduled for certification in April 1997, and one in June 1997. Four other transit agency plans were scheduled to be reviewed in 1997, along with the Washington State Transportation Plan.

9   This problem may reflect the fact that the mandate to designate critical areas and resource lands does not provide a clear linkage to other comprehensive plan requirements spelled out in the Growth Management Act. According to state planners, the designation and protection of environmentally significant lands was mandated to be done in the first six months of the planning process in order to avoid a rush on applications from the building industry to develop in critically sensitive areas before more stringent policies and laws came on the books.

References

Bay Area Council (1988) *Making Sense of the Region's Growth*, Bay Area Council, San Francisco.

Bollens, S. (1992) State Growth Management/Intergovernmental Frameworks and Policy Objectives, *Journal of the American Planning Association*, Vol. 58, No. 4, Autumn, pp.454-466.

Bosselman, F., and D. Callies (1972) *The Quiet Revolution in Land Use Control* (prepared for the Council of Environmental Quality) United States Government Printing Office, Washington, DC.

Brower, D., D. Godschalk and D. Porter (eds) (1989) *Understanding Growth Management/Critical Issues and a Research Agenda*, Urban Land Institute, Washington, DC.

Deakin, E. (1989) Growth Controls and Growth Management: A Summary of Review of Empirical Research, in D. Brower, D. Godschalk and D. Porter (eds) *Understanding Growth Management*, Urban Land Institute, Washington DC, pp. 3-21.

DeGrove, J., (ed.) (1991) *Balanced Growth/A Planning Guide for Local Government*, International City Management Association, Washington, DC.

DeGrove, J. (1992) *The New Frontier for Land Policy/Planning and Growth Management in the States*, Lincoln Institute of Land Policy, Cambridge, MA.

DeGrove, J., and N. Stroud (1987) State Land Planning and Regulation: Innovative Roles in the 1980s and Beyond, *Land Use Law and Zoning Digest*, Vol. 39, No. 3, March, pp. 3-8.

Denver Regional Council of Governments (1996) *Metro Vision 2020* (draft), Denver Regional Council of Governments, Denver, CO.

Detwiler, P. (1992) Is Cooperation Enough? A Review of San Diego's Latest Growth Management Program, in D. Porter (ed.) *State and Regional Initiatives for Managing Development/Policy Issues and Practical Concerns*, Urban Land Institute, Washington DC, pp. 57-80.

DiMento, J., and L. Grayer (eds) (1991) *Confronting Regional Challenges/Approaches to LULUs, Growth, and Other Vexing Governance Problems*, Lincoln Institute of Land Policy, Cambridge, MA.

Epling, J. (1992) Growth Management in New Jersey: An Update, *Environmental and Urban Issues*, Vol. 17, No. 4, Summer, pp. 6-16.

Gale, D. (1992) Eight State-Sponsored Growth Management Programs/A Comparative Analysis, *Journal of the American Planning Association*, Vol. 58, No. 4, Autumn, pp. 425-439.

Godschalk, D. (1986) Urban Development, in F. So, I. Hand and B. McDowell (eds) *The Practice of State and Regional Planning*, American Planning Association, Chicago, IL, pp. 309-339.

Godschalk, D., and D. Brower (1989) A Coordinated Growth Management Research Strategy, in D. Brower, D. Godschalk and D. Porter (eds) *Understanding Growth Management*, Urban Land Institute, Washington DC, pp. 159-179.

Johnson, R., S. Schwarz and S. Tracy (1984) Growth Phasing and Resistance to Infill Development in Sacramento County, *Journal of the American Planning Association*, Vol. 50, No. 4, Autumn, pp. 434-446.

Kasowski, K. (1992) The Costs of Sprawl, Revisited, *National Growth Management Leadership Project*, Vol. 3, No. 2, September, 1000 Friends of Oregon.

Keefe, S. (1992) Twin Cities Federalism: The Politics of Metropolitan Governance, in D. Porter (ed.) *State and Regional Initiatives for Managing Development/Policy Issues and Practical Concerns*, Urban Land Institute, Washington DC, pp. 81-142.

King County (1985) *King County Comprehensive Plan*, King County Department of Parks, Planning and Resources, Seattle, WA.

Local Governance Study Commission, Washington State (1988) *A History of Washington's Local Governments*, Vol. I, Institute for Public Policy, The Evergreen State College, Olympia, WA.

McDowell, B. (1986) The Evolution of American Planning, in F. So, I. Hand and B. McDowell (eds) *The Practice of State and Regional Planning*, American Planning Association, Chicago, IL, pp. 23-62.

Metropolitan Council in the Twin Cities Area (1986) *Metropolitan Development and Investment Framework*, Metropolitan Council in the Twin Cities Area, St. Paul, MN.

Perin, C. (1977) *Everything in Its Place: Social order and land use in America*, Princeton University Press, Princeton, NJ.

Piro, R. (1993) *Growth Management in an Urban Regional Context: the Contemporary Transformation of Regional Development Planning from a Governance Perspective* (unpublished doctoral dissertation), University of Washington, Seattle, WA.

Porter, D., (ed.) (1992) *States and Regional Initiatives for Managing Development/Policy Issues and Practical Concerns*, Urban Land Institute, Washington, DC.

Puget Sound Regional Council (1995a) *The Metropolitan Transportation Plan*, Puget Sound Regional Council, Seattle, WA.

Puget Sound Regional Council (1995b) *VISION 2020 Update*, Puget Sound Regional Council, Seattle, WA.

Rapaport, R. (1992a) Regionalism: Part 1 – The Failed Dream, *San Francisco Focus*, January, pp. 46-49, 82-88.

Rapaport, R. (1992b) Regionalism: Part 2 – A New Approach, *San Francisco Focus*, February, p. 77, 98-103.

San Diego Association of Governments (1992) *Regional Growth Management Strategy* (Revised Draft), San Diego Association of Governments, San Diego, CA.

Schaenman, P., and T. Muller (1975) Land Development: Measuring the Impacts, in Scott et al. (eds) *Management and Control of Growth*, Vol. II, Urban Land Institute, Washington DC, pp. 494-500.

Scott, R., D. Brower and D. Miner (eds) (1975) *Management and Control of Growth/Issues, Techniques, Problems, Trends*, Vol. I-III, Urban Land Institute, Washington, DC.

So, F., I. Hand and B. McDowell (eds) (1986) *The Practice of State and Regional Planning*, American Planning Association, Chicago, IL.

Washington, State of (1994) *Revised Code of Washington*, 1994 Edition, State of Washington, Olympia, WA.

Washington State Department of Transportation (1989) *Statewide Land Use and Transportation Planning Survey Results and Analysis*, Report RD-89-05, June, Washington State Department of Transportation, Olympia, WA.

Chapter 8

ROM-Rijnmond:
Programme Management in Practice

A. Dijkstra[1]

8.1   Introduction: Background to the ROM-Rijnmond Project

In the Fourth Report on Spatial Planning (1988) and the National Environmental Policy Plan (1989), the national government in the Netherlands decided to take initiative in establishing integral policy plans for certain regions. These plans are meant to ensure that the interests of the environment play a more active role in decisions about spatial use. Eleven 'ROM' regions were introduced (ROM being the Dutch abbreviation for Spatial Planning and Environment). One of the ROM regions is the Rijnmond region.

At that time, the Rijnmond region was on the threshold of a number of far-reaching developments and many choices had to be made. The strain on the environment was still far too great in this region despite the considerable efforts which had already been made. Spatial quality as well as environmental quality is subject to both threats and opportunities. There was a demand for sufficient high-quality housing, nature reserves and recreational facilities.

Furthermore, the port of Rotterdam had to be prepared to maintain and further develop its leading role in the world. This meant staying one step ahead of the competition by improving the quality of existing activities and stimulating and developing new activities. These challenges led to the definition of a dual objective for this region: the expansion of the main port in Rotterdam and the improvement of the living and social climate in the region.

Three sectional documents were published on the problem areas; the Main Port Target Scenario, the Spatial Arrangement Target Scenario and the Environmental Objectives Report. The wishes and goals for this region up to the year 2010 are recorded in these documents. An analysis was carried out using these target plans and objectives as a basis for determining which favourable development opportunities were feasible and for gaining insight into which bottleneck problems might have to be solved in order to see these opportunities realised. On the basis of this analysis, the parties determined that the realisation of a dual objective for this region was feasible. It would require some extra effort in addition to the unimpaired implementation of existing policy. Those extra efforts were crystallised into a large number of concrete projects. These projects are described in the ROM-Rijnmond Plan of Approach.

When the ROM-Rijnmond Policy Covenant was signed in December 1993, it was agreed that the ROM-Rijnmond Plan of Approach would be carried out collectively. The covenant also included stipulations to ensure progress checks on and monitoring, evaluation and re-verification of the projects. A total of 23 parties, ranging from local government and industry groups to the national government, put their signatures to the approach to the problems that had been identified in the Rijnmond region.

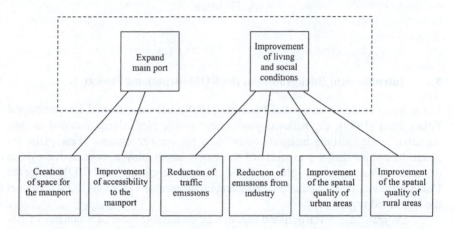

Figure 8.1   Sub-objectives that contribute to the realisation of the dual objective

The heart of the Policy Covenant is formed by the dual objective, which expresses the target scenarios for the period up to 2010. The parties all aspire towards a stronger economy and a better environment. Rotterdam is a world port and it needs to remain so. That is why the position of Rotterdam as a port must be enhanced through innovation, an increase in accessibility and possible harbour expansions. At the same time, the region must not just continue to be inhabitable, but the quality of the living environment must be improved. This can be achieved by promoting enlargement cases such as the Regional Green Plan and through measures to prevent disruptions to the environment.

The sub-objectives were in turn crystallised into 26 desirable results that are quantifiable and must be achieved by the year 2010. Some examples of the results listed include a reduction in traffic congestion and NOx exhaust, as well as measures to reduce the number of residences subject to noise pollution and an increase in the total number of acres that can be used for industrial parks. In an effort to realise these goals, a great many operations had already been set in motion. However, these were not sufficient for reaching the designated objectives. So in addition to the activities that occurred outside of the ROM-Rijnmond framework, 47 extra projects were defined, totalling an estimated cost of four billion euros (about four billion USD). Some of the noteworthy projects are a study on the laying of a second *Maasvlakte*

(Meuse-river plain) along the coast, improvements to the 'Maricor' (the Maasvlakte-Ridderkerk corridor) and its most important traffic artery, the A15, and the re-development of old harbour areas in Vlaardingen and Schiedam.

8.2　From Project to Programme

In the period before the Plan of Approach was drawn up, a lot of time went into reaching agreement between the 23 parties, each with their different interests. A consensus of opinion had to be reached concerning the crystallisation of what kind of development was desirable in the region and what means should be used to determine its essence. A description of the guidelines of the projects including an approximate appropriation of costs and responsibilities was also necessary.

During the organisational phase, ROM-Rijnmond had all the characteristics of a project, with a beginning (Start Covenant), an end (Policy Covenant) and a definite result (Plan of Approach). This project was determined by the factors time, money, quality, information and organisation. The nature of the plan changed somewhat during the finalisation process. The parties had to deliver concrete contributions in order to realise the objectives. People wanted to be able to measure whether what was going to be achieved would indeed have the desired effect. This was not a loose collection of projects, but an integrated relationship between an intricately connected set of goals and activities. That meant that direction at project level or multi-project management would not suffice. There was a clear need for direction at a higher, more abstract level: that of programme management.

Programme management is similar to project management, but there are a few important differences. Whereas in a project people work towards clearly described end results that have been stipulated ahead of time, a programme can be more vague. The definition of a programme is as follows: a coherent collection of important, goal-oriented efforts. These efforts could be in the form of result-oriented one-off projects, the overall results of which are unforeseeable. A programme is concerned with implementing a certain policy that has both quantifiable elements and more ephemeral elements. In general, programmes also run for a longer period of time and are more flexible. Objectives can be altered and other means applied. The programme creates order in the midst of a large number of independent actions. A few factors are crucial to the successful running of a programme: an adequate implementation organisation; cohesion within the programme; and progress controls. These three things are addressed in turn below.

8.3　The Implementation Organisation

It is important for the various parties involved to come together on a regular basis to discuss the overall programme, the individual projects and associated matters. The parties who signed the ROM-Rijnmond Policy Covenant all work together in the ROM-Rijnmond Implementation Organisation. The Programme Bureau is part of the

Implementation Organisation. It co-ordinates and facilitates the programme's implementation but is not responsible for the manner in which the individual projects within the programme are carried out. An external consultant is hired to act as the programme manager. This external advisor must make sure that the political and managerial negotiations go smoothly and is responsible for consolidating the mutual rapport between the parties.

The Implementation Organisation also consists of: a ROM-Rijnmond Administrative Consultation Board (abbreviated in Dutch to BOR); a ROM-Rijnmond Daily Management Department (abbreviated in Dutch to dbBOR); and a ROM-Rijnmond Consultation/Co-ordination Group (abbreviated in Dutch to OCR).

The Administrative Consultation Board is the highest umbrella organisation in the ROM implementation organisation. It is made up of the three ministries, the county council of Zuid-Holland, the City of Rotterdam and the Rotterdam Urban Area Council (representing 14 cities) and organised business. The Administrative Consultation Board oversees the execution of the Plan of Approach and makes the important decisions. The Daily Management Department has the same basic nature as the Administrative Consultation Board but concentrates more on short-term issues, those things that cannot wait until the Administrative Consultation Board meeting occurs. In addition, the Daily Management Department determines whether sufficient and valuable information is on hand for the meetings of the Administrative Consultation Board. The Consultation/Co-ordination Group is the official forum by which the parties involved are represented at the civil and directorial level. Here, subjects are discussed and abridged ahead of time so that they encompass only the essential administrative issues.

Each party in the covenant is a commissioning agent and, as such, initiator of one or more of the 47 projects. Thus, each party provides project leaders for the ROM projects. The project leader follows the usual method of his/her agency and reports initially to the commissioning agent in his/her own organisation. Project leaders make quarterly reports to the Implementation Organisation, which oversees the progress and bottlenecks of each particular project and the interrelation to other projects. The Implementation Organisation fulfils a purely facilitating role and is not able to levy any sanctions in the case of defaults. The parties, though, can call one other to account through the different consultation bodies.

In such an implementation organisation, good communication is very important. The foundation both within ROM-Rijnmond and outside of it is crucial to the continued progress of the programme and the fulfilment of the dual objective. An important aspect of ROM-Rijnmond is its collective recognition of the region's problematic nature and the concerted effort to find solutions. Communication plays a big role in gaining and maintaining this collective base. The communication is primarily aimed at internal organs and not necessarily intended for the public. ROM-Rijnmond's main message is:

> The parties united in ROM-Rijnmond are making a huge effort to expand the main port and improve the living and social climate. This is being achieved through a dynamic programme of projects. The collaboration between all parties involved increases the feasibility of all of the projects.

One important means for publicising this message is the bi-monthly publication *ROM Active*, which addresses special themes related to ROM-Rijnmond, such as train transportation or energy use. It also devotes attention to the results being achieved within various ROM-Rijnmond projects.

8.4 Cohesion within the ROM-Rijnmond Programme

Another important aspect to the programme is its ability to steer towards different objectives that are mutually connected without having to achieve all the objectives in every project or activity. For instance, when a road is built it might not be feasible to compensate for the damage done to the environment in that project, but compensation can be made elsewhere within the programme. A number of instruments have been developed to provide some insight into these programme interconnections.

To clearly define the connections between the dual objectives, the subobjectives, the desired results and the projects, an *Objective-Means-Hierarchy* was developed. The biggest advantage of this schematic is that it shows at a glance what the connections are between the different parts of the programme.

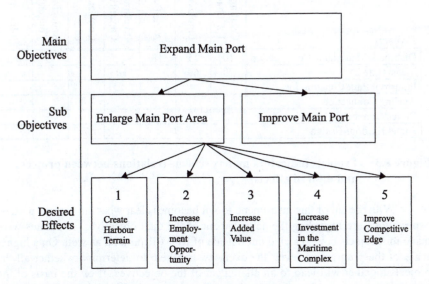

Figure 8.2 Objective-means hierarchy (fragment)

The Objective-Means Hierarchy was drawn up as follows. Six sub-objectives were derived from the main objectives of expanding the main port and improving the living and social climate. These sub-objectives were detailed in quantifiable, desired results. Then, the 47 ROM-Rijnmond projects, all of which make contributions, were recorded as illustrated by Figure 8.2. The fact that the connections between one project and another and between the projects and the desired results are made clear encourages the individual project leaders to feel more strongly committed to the total programme and to be aware of more than just their own project.

The second instrument to be developed was a *Project-Indicator-Matrix*. Indicators were defined for each of the desired results. In the Project Indicator Matrix illustrated by Figure 8.3, the projects and these indicators are mapped out in relation to one another. Estimates are made for each project, regarding how much that project contributes, positively or negatively, to the achievement of the desired results. For example, the Second Maasvlakte project is expected to create 70 per cent of the total expansion area planned in the Plan of Approach for industrial parks and harbours. The project is also expected to help increase job opportunities in the area. However, this will in turn attract more traffic and consequently have a negative effect on the desired 'reduction in traffic congestion'.

	Desired effect						
	1	2	3	4	5	6	7
Projects	Creating 1,270ha	Employment	Added Value	Investment Capacity	Compet. edge	Reducing Congestion	Infra. effic.
Distripark Maasvlakte Construction	10	15	18			-5	pm
Maasvlakte 2	70	60		10			
Industrial Park Construction							
Aveling Connection							
Botlek Connection							
Dintel Harbour Bridge					7		

Figure 8.3   Projects indicators matrix scheme (relations between projects and desired effects)

With the aid of this instrument, too, it becomes clear what the various parties' expectations are regarding the achievement of the dual objectives. Estimates were made for projects being carried out outside of ROM-Rijnmond as well. On a higher scale, at the programme level, the overview can help to determine whether all the projects together will lead to an attainment of the objectives. If on the basis of the estimates it becomes clear that certain results are unattainable, a collective decision can be made as to which projects should be bolstered and which should be shut down.

This instrument is not only useful for clarifying the connections between projects, but it can also help register the progress being made towards the attainment of the desired results. If circumstances change along the way, it naturally has certain consequences: projects stagnate or fail to contribute to the desired course of action. It may become evident at an interim stage that certain objectives are not going to be fully met. When this happens, it is possible to re-direct matters. Projects can be given priority or stopped altogether, or new projects can be started up depending on the precise situation.

8.5 Progress Controls

Adequate monitoring of the projects is crucial to the programme's success. The parties involved need to have access to reliable, recognisable progress reports, and the methodology used should be such that the parties can be confident of the results. The outcome of the progress reports forms the basis for the collective management of the programme.

Three different areas of information must be closely monitored in order to manage the ROM-Rijnmond programme effectively: the progress of the projects; developments in the surrounding area; and the monitoring of previously selected indicators.

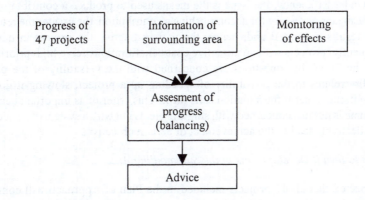

Figure 8.4   The three areas of information evaluated by the Programme Bureau

Data on the Progress of the Projects

Every quarter, discussions are held with the project leaders of the ROM-Rijnmond projects. Among the matters covered in these discussions are planning, financing, organisation and communication. This is the forum for reporting any problems to the implementation organisation. Wherever possible, the implementation organisation tries to find solutions to those problems.

Once a year, the progress of all 47 ROM-Rijnmond projects is carefully evaluated. In the beginning this meant that each project was assessed according to the 'THEFD' criteria. THEFD is the Dutch acronym for Tempo, Feasibility, Efficiency, Flexibility and Goal Orientation. Not all the information that was collected was necessary to steer the programme. So it was decided to collect only the information on Tempo, Feasibility and Goal Orientation. The data for this evaluation are collected from questionnaires filled out by a number of people directly involved in the projects. The people questioned give a relative score to each question.

Under the Tempo heading, the issue is whether the project delivers a valuable contribution to the desired effects in a timely manner. The sooner a certain contribution is made, the more attractive the project is, because it leads to a faster realisation of the desired effects. The questions under Feasibility concern the probability of the project actually furnishing the intended contribution to the dual objective. Projects that have a high certainty rate of supporting the effectuation of the objectives and a low risk of failure are given a high priority.

Finally, questions are posed about Goal Orientation: to what extent does the project contribute to the objectives that have been set? Projects with the greatest level of goal orientation are naturally given priority. The extent of goal orientation, though, can change suddenly according to circumstances.

These data are recorded for each project. Should problems be encountered on the basis of the data and the conclusion is drawn that some adjustments should be taken then the Efficiency and Flexibility criteria are also examined.

Under Efficiency, the issue is the means used to produce a contribution. The fewer the means necessary for accomplishing a contribution, the greater the returns or desired results generated and, hence, the more attractive the project in question. Consequently, projects with an attractive cost/benefit ratio receive high priority.

The Flexibility questions are concerned with the versatility of the project. Versatility relates to the possibility of speeding up a project, slowing it down or stopping it altogether if the situation calls for it. This criterion is important because a programme is not just concerned with single projects but with a system of projects that must collectively lead to the achievement of certain objectives.

Data on Relevant Developments in the Surrounding Area

It is expected that all 47 projects included in the Plan of Approach will contribute directly to the realisation of the 26 desired results. Another substantial contribution, however, will be provided by projects outside the ROM framework and by autonomous developments.

It is therefore important to gather information on relevant developments in related areas which might influence the main port and the living and social climate in the Rijnmond region and, accordingly, the ROM-programme. To this end an annual reconnaissance is made of regular projects in the surrounding area that are not included in the Plan of Approach but are helping to attain the desired effect. These projects are called the 47+ projects. Examples include the construction of the Betuwe train line, the building of the second Benelux tunnel, the Project for the Reduction of Industrial Noise Pollution, Carbohydrates 2000, the development of woodland within the context of the Randstad Green Project, the Voordelta Action Programme and the Industrial Target Groups Policy.

The exploration of surrounding areas is not just limited to relevant projects in the immediate area. The Rijnmond region is part of a greater whole, and the development of its economy, environment, spatial planning and traffic and transportation are all influenced by many factors. For this reason, an annual trend analysis is also made of the most important developments in the vicinity.

The Monitoring of the Results of the Previously Selected Indicators

In addition to following the programme, it is important to investigate annually how many results designated in the programme are actually being attained. For this reason a monitoring system (which could be compared to a thermometer) was developed to easily show at any given moment to what extent the desired effect has been reached and in what measure the different projects contribute to its realisation.

The monitoring system is an important tool in drawing up progress reports: it details each of the 26 desired effects. A good set-up for such a monitor is essential since work has to be carried out according to that model for the entire duration of the programme. It needs to be a monitor that is familiar to all parties and as such, methods of measurement must be used that produce results that everyone can believe in.

8.6   Balancing the Three Different Areas of Development

Information that has been collected from the three different areas of development must be considered in a mutual context. These three facets are the ROM projects, projects and relevant developments in the Rijnmond region outside the ROM framework, and the current situation of the desired effects. *Balancing* refers to the process of assembling the three areas of information, looking at them as a cohesive whole, and analysing them. It is called balancing because on the one hand the pros and cons are weighed against each other and on the other a balance is found between both elements of the dual objective. This concerns interconnections and relationships between the different activities, i.e. projects, and the influence of topical external developments on the programme. It is essential in the balancing process that the three areas contain information that is current, precise and reliable.

Important issues in balancing are: to what extent is the realisation of the dual objective getting closer?; how do the efforts and results relate to one another?; and what are the influences of autonomous projects and topical external macro-factors?

The Programme Bureau analyses the data at four levels, those used in the Objective-Means Hierarchy. These levels are: the project level (for the 47 ROM projects and the 47+ projects); the desired results level; the sub-objectives level; and the dual objective level, that is the total programme level.

Analysis at Project Level

On the basis of the questionnaires, the value of the operating criteria is recorded for each ROM-Rijnmond project. The information provided is intended for use in plotting project developments over time and relating particular information on any one project to the other projects.

For the 47+ projects, the analysis is aimed at the Tempo and Goal Orientation operating criteria. On the basis of this information, it becomes clear whether the projects deviate greatly from the expectations expressed in the ROM-Rijnmond Plan of Approach.

Analysis at the Desired Effect Level

The next level to be analysed is that of the desired effect. For each desired result, a plot is made of which projects (ROM and 47+) contribute to that effect and what the monitoring results have shown (e.g. what are the levels of NOx emissions in the area?). In other words, this is the analysis level where the three areas of information, projects, surroundings and monitoring, come together.

Analysis at the Sub-objective Level

The analysis at the sub-objective level is built up from the results of the analysis at the desired effect level. Under each sub-objective, there are a number of desired results that contribute to it. How the desired effects develop is then indicated per sub-objective. For instance, the sub-objective to 'enhance the spatial quality of the urban area' is built up from two desired effects: improvements in the spatial quality of both the harbour-related work areas and the residential areas around the main port. An aggregation of the information on these two desired effects is used to help estimate how the sub-objective will develop.

Analysis at the Dual Objective Level

Based on the results from the analysis at the sub-objective level, an analysis is made of the dual objective level. This consists of a summarised account of the results and developments from the six sub-objectives. This aggregation provides insight into how close a balanced realisation of the dual objective really is.

Based on the results of analysis done at four levels, the Programme Bureau can indicate what if any action needs to be taken to redirect the ROM-Rijnmond Programme. The Programme Bureau might, for example, propose that projects be initiated, implemented or even terminated after realisation. It might also suggest that extra attention be paid to one particular project.

Once every four years, the Plan of Approach is evaluated and re-verified as necessary. The information provided by the balancing process can then be used to adjust the desired effects and, subsequently, the objectives.

8.7   Conclusions: The First Results of ROM-Rijnmond

In the first three years, the process was thoroughly organised and the train set in motion. The ROM approach was developed, tested and shown to work. Though initially the projects tended to get no further than good intentions, the first results are showing. Although the desired effects are still far away. A number of important projects were initiated up in the context of the ROM framework. By steering at a higher abstract level, the programme management methodology has managed to book the first results for ROM-Rijnmond towards ultimately attaining the objectives. It is important to keep in mind that the ROM-Rijnmond programme fulfils a supplementary role. Many results fall outside of the programme. One characteristic of a supplementary programme is that this is where all the problematic cases end up. It is difficult to determine which part of the results are on account of ROM-Rijnmond-projects and which part are results of the developments in the surrounding area. One other problem is how to collect recent input for the monitoring.

The programme functions in a dynamic environment. Traffic continues to increase, as does the accompanying environmental disruption. Projects meant to combat this disruption are sometimes not very effective. Other projects for resolving traffic congestion and impact, such as the various infrastructure projects, got off to a promising start, but the needs for the new infrastructure are outgrowing those projects faster than the tempo with which they can reasonably be expected to be financed and realised. In addition, it takes time to implement measures and the effects of these measures is not always immediately noticeable. The programme management system identifies such problems and brings them under discussion.

A major function of the ROM-Rijnmond-programme is as an important means of communication. Within the ROM-Rijnmond framework, people know where to find one another and commitment can often be gained thanks to the agreements made ahead of time. The consultation committee provides the opportunity for discussing and solving procedural obstacles, and the existence of a relatively independent secretariat guarantees a certain level of orderliness. Collaboration constantly demands a great degree of openness and transparency from people. The programme management system offers a good instrument for following developments in the region as objectively as possible and for making specific suggestions for re-direction. The system also offers good possibilities for ensuring that the agreements are made in a timely manner.

Note

1   A. Dijkstra is working as a spatial and environmental planner at the Environmental Department of the Directorate of Public Works at the City of Rotterdam. Since 1995 she is detached to the Program Office ROM-Rijnmond, to contribute to the program management of ROM-Rijnmond and is co-ordinator of the ROM-Rijnmond projects run by the City of Rotterdam.

References

Engels, M., and A. Duijts (1996) *ROM Programma-management*, Programmabureau ROM-Rijnmond, Rotterdam, The Netherlands.
Doorewaard, M.E.M., and J.W.D. Peters (1995) *Management Consult in Bedrijf, III.3, Doelgericht managen van complexe programma's*, Management Consult Kluwer Bedrijfswetenschappen, Deventer, The Netherlands.

Chapter 9

Environmental Enhancement by the Environment Agency in South East England[1]

H.R. Howes[2]

9.1   Introduction

The new Environment Agency for England and Wales is one of the most powerful environmental regulators in the world. It provides a comprehensive approach to the protection and management of the environment by combining the regulation of land, air and water.

The Agency came into being on the 1st April 1996. It has been created out of the National Rivers Authority (NRA), Her Majesty's Inspectorate of Pollution and from the waste regulatory function of the County Councils and Waste Regulation Authorities. It is therefore able to take a much wider view of environmental regulation than was possible for its predecessors.

The main strands of the Agency's powers and duties are water management and the prevention and control of pollution. Examples are regulating those industrial processes with the greatest polluting potential of both air and water, securing the proper use of water resources, an improving the quality of rivers, estuaries, and coastal water.

The Agency is a statutory consultee in the Town and Country Planning System. It has the right to be consulted on development plans and on a wide range of planning applications. It is particularly concerned to have a major influence on the review of the regional planning guidance for the south east by building on the success of the former NRA in securing advice to local planning authorities on rivers, estuaries, the coast and marine environment, water supply and waste disposal.

9.2   Thames Region

Thames region comprises the counties of Berkshire, Hertfordshire, Oxfordshire and Greater London, the major parts of Buckinghamshire and Surrey and parts of Bedfordshire and Essex, Gloucestershire, Hampshire and Wiltshire.

The region is home to 12 million people which amounts to over 23 per cent of the population of England and Wales. It covers some 1,302,400 hectares which is 8.6 per cent of the area of England and Wales. It contains 24.5 per cent of all households.

This makes it simultaneously both the most populous and geographically the second smallest of the Environment Agency regions. The density of the population is 9.1 people per hectare. This compares with a figure of 4.9 people per hectare for north west region which is the next most populous.

In economic terms Thames region enjoys pre-eminence in information technology and electronics and biotechnology. On average for 1991-95 26 per cent of the value of contractors new orders for construction takes place in the Thames region. This gives rise to 26 per cent of total planning applications in England and Wales. There is a working population of 6.4 million of which 88 per cent are in the service sector. This is higher than for any other Environment Agency region. The region's share of gross domestic product amounts to 27 per cent. This corresponds to GDP per head of population of £11,600 in contrast to the next highest of £10,800 in the Southern region.

In geological terms Thames region covers a significant part of the English scarplands which contribute to the attractiveness and diversity of the region's environment. The region corresponds to the basin of the River Thames and its tributaries. The main elements are the Upper Thames basin lying between the Cotswold Hills and the Chiltern Hills, the middle Thames basin lying in the south east of the Chilterns, the Lee Valley and its tributaries and the London conurbation. The government office for the south east has identified as a key issue the protection and enhancement of the environment while achieving continued economic growth to which tourism is seen as a major contributor. The region supports a high quality environment. Over 40% of the region is classified as Areas of Outstanding Natural Beauty, Green Belt or some other form of protection.

9.3   Thames 21 – A Strategic Approach to River Basin Management

In 1995 the NRA published *Thames 21 – A Planning Perspective and a Sustainable Strategy for the Thames Region.* It provides principles of sustainable development in relation to all the NRAs core functions. It also provides analysis of the region's major growth locations in terms of the key catchment planning issues for inclusion both by the NRA in its catchment management plans and by local planning authorities of their development plans.

Thames 21 has three roles. First, it acts as a bridge with external organisations dealing with strategic planning. In this role it articulates water related interests in regional planning guidance. Second, it provides any easy-to-use summary of NRA policies for promotion through the statutory development plan system. And third, it provides a regional context for the preparation of catchment management plans with an indication of the development issues which these plans will need to address.

Thames Environment 21: the Environment Agency's perspective for land use planning in the Thames Region is a new initiative for the Environment Agency. It provides guidance to local planning authorities on the comprehensive environmental strategy which the Agency proposes to follow in the Thames Region. It is specifically intended to assist the regional planning conferences (SERPLAN and the South West

Regional Planning Conference) in revising the Regional Strategies. As a statutory consultee in the Town and Country Planning system, the Environment Agency places great importance on securing appropriate regional planning advice for the environment. It is also intended to provide local planning authorities with the full range of environmental topics in development plans.

Thames Environment 21 fulfils the requirements on the Environment Agency to make a contribution towards achieving sustainable development. It therefore provides a basis for both analysing the impact on the environment of major development locations and for assessing the merits of alternative locations for development. By indicating the Environment Agency's general requirements for environmental enhancements and mitigations well in advance of development taking place the Agency feels confident that a high standard of development may be achieved than might otherwise be the case.

9.4 Sustainable Development and Costs and Benefits

Sustainable Development

The Environment Agency has a statutory duty to protect and enhance the environment taken as a whole so as to make a positive contribution towards achieving sustainable development. In dealing with development the Environment Agency will have regard to its twin duties under the Environment Act:

Section 4:   A principal aim of the Agency is discharging its functions to make a positive contribution towards attaining the objective of achieving sustainable development.

Section 39: General duty to have regard to cost and benefits in exercising its powers.

Sustainability is sometimes seen as being opposed to development. However, the Agency considers that development can generally be achieved in ways that are compatible with environmental protection and enhancement. In particular high environmental quality is crucial to achieving economic growth and regeneration and a better quality of life for the region's population. There are also opportunities to maximise environmental benefits from redevelopment schemes. The development of 'brownfield' sites in the Thames Gateway is a good example. Our approach is therefore concerned with quality and quantity of development and its general appropriateness in a particular location.

The guidance from the Department of the Environment recognises that sustainable development involves reconciling the need for economic development with the protecting and enhancing of the environment without compromising the ability of future generations to meet their own needs. The Agency is to take account of all types of costs and benefits when making such decisions. This will ensure that financial and other considerations are taken into account, but also that environmental considerations are given the central role that is necessary for sustainable development. Principles of sustainable development in the Thames Region include the following.

- Achieve a continuing overall improvement in the quality of surface and groundwaters through the control of pollution, enforcement of the polluter pays principle, and avoidance of irreversible impacts upon water quality.
- Manage areas at risk from flooding for the benefit of people, property and the environment.
- Achieve sustainable waste management through implementation of the waste hierarchy and proximity principles.
- Protect, improve and promote the water environment and waterside land for the purposes of amenity and appropriate recreational use.
- Minimise pollution levels wherever possible through use of best available technology not involving excessive costs and best practicable environmental option.
- Conserve and enhance the water environment, including the aquatic and terrestrial ecology, landscape and archaeological features.
- Manage ground and surface water resources to achieve the right balance between the needs of society and the environment.

Costs and Benefits

In its relationship with the development industry the Agency is required to be 'reasonable' in terms of the environmental enhancements it may require of developers. 'Reasonable' is not defined. It is clearly 'reasonable' to ask for substantial enhancements if these are financed out of the gratuitous increase of the site value brought about by the granting of an outline planning permission, Enhancements of a lesser order would be 'reasonable' when a developer, having acquired the site, seeks detailed planning permission for the scheme.

Figure 9.1a shows how the developer bases his calculation on the market value of the completed development. From this he subtracts the development costs, finance costs and an element of profit. Unless he is aware of a requirement for environmental enhancement the residual sum will be the maximum he is prepared to pay for the site. Clearly it is too late at this stage for the Environment Agency to require a substantial mitigation package.

Figure 9.1b demonstrates what happens when the Environment Agency specifies its requirements for environmental enhancements in advance preferably through the local planning authority's development plan, through a planning brief, or through a planning agreement between the land owner and a local planning authority. In this case the developer takes the environmental enhancements into his financial calculations and offers correspondingly less for the site. The land owner clearly will make less profit on the sale but as the environmental enhancements are the key to his releasing the development value of the land he is still better off than if he had held the land back.

Figure 9.1c shows that the environmental enhancements are likely to increase the value of the completed development and they are likely to go a long way towards financing themselves.

Figure 9.1d reflects the situation where planning policies are seeking to divert development pressures to redevelopment of brownfield sites. The value of the development to the purchaser will not be greater than if the development had taken place on a greenfield site. Furthermore there are additional site costs to be taken into account and there may therefore be less scope for the Environment Agency to seek substantial environmental enhancements. Nevertheless there are many examples of successful enhancements being included in development schemes particularly in the case of retail planning permissions which result in a very large increase in site value.

9.5 Development Pressures in South East England

The Government's recent Green Paper *Household Growth: Where shall we live?* indicates that some 4.4 million houses will be required nationally in the period 1991-2016. This is likely to mean an additional 10,000 houses per annum in the South East which would impact on the Agency's concerns related to water resources, flood plain policy, conservation, waste, water quality, air quality and aggregates.

Water Resources

The level of development proposed in the Green Paper is likely to give rise to an increase in the demand for water of approximately 5 per cent. If it proves impossible to use water more efficiently it is likely that the need for new water resource schemes could be brought forward to the pre-2016 period. This is not a long enough time horizon for the implementation of a major project and clearly should be a material consideration in the strategy for the Region.

However the increase in demand could be balanced out by the implementation of a range of water efficiency measures. The Agency will use the opportunity of the review of the Regional Planning Guidance to promote such measures. Demands and resources are being reviewed at the present time as part of the asset management plan process and a clearer picture is likely to emerge during 1998.

Floodplain Protection

Flooding of a floodplain is a natural and desirable phenomenon where it can occur without risk to human life. However the ability of floodplains to function in this way has been eroded over the years as the land in them has been developed. Consequently, the channels and floodplains of many major rivers have become restricted, especially in urban areas. Inevitably, these restricted channels could not accommodate large storm flows and serious flooding of developed areas has occurred. In coastal areas, the loss of large areas of floodplain has placed people and property at risk from flooding and led to pressure for new or improved coastal defences.

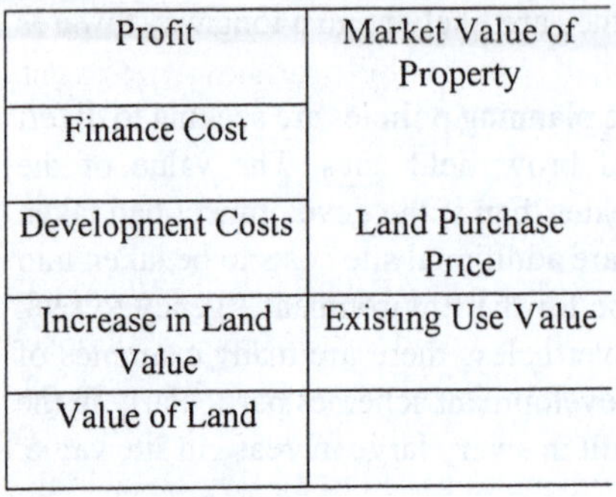

Profit	Market Value of Property
Finance Cost	
Development Costs	Land Purchase Price
Increase in Land Value	Existing Use Value
Value of Land	

Figure 9.1a

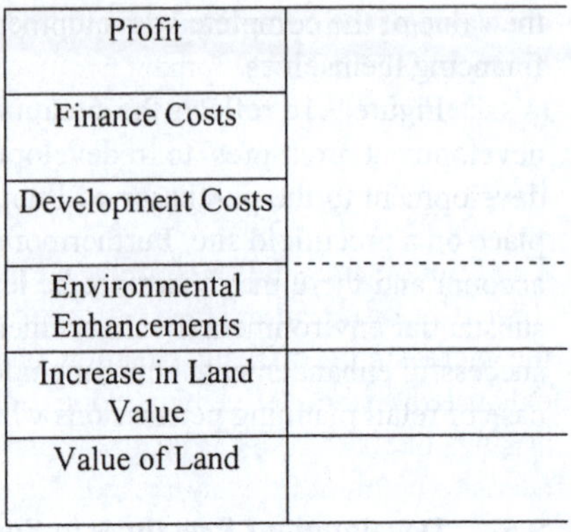

Profit	
Finance Costs	
Development Costs	
Environmental Enhancements	
Increase in Land Value	
Value of Land	

Figure 9.1b

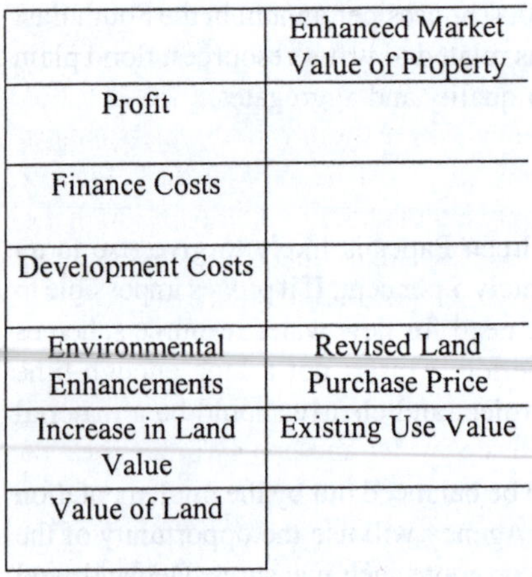

	Enhanced Market Value of Property
Profit	
Finance Costs	
Development Costs	
Environmental Enhancements	Revised Land Purchase Price
Increase in Land Value	Existing Use Value
Value of Land	

Figure 9.1c

Profit	Market Value of Property
Finance Costs	
Development Costs	
Demolition and Site Clearance Costs	
Environmental Enhancements	Revised Land Purchase Price
	Existing Use Value
Value of Land	

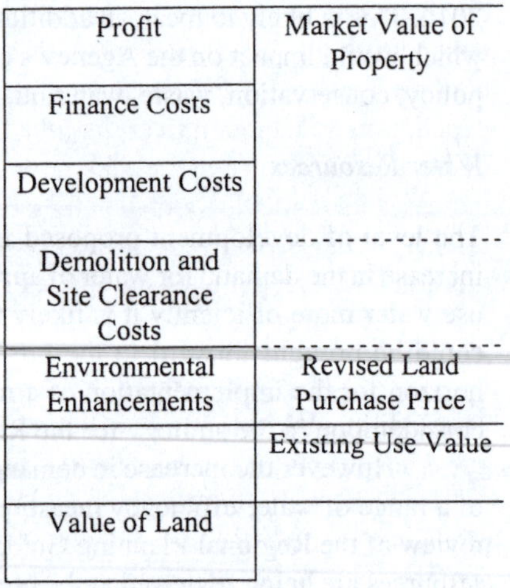

Figure 9.1d

Figure 9.1　Costs and benefits; the developer's calculations

Conservation

Some parts of the south east may have already been developed beyond the capacity of the environment to sustain the existing uses. For instance this may be due to insufficient clean water resources, insufficient clean air, or general environmental degradation due to lack of greenspace and wildlife. Such areas should not be further degraded as this will lead to reduced economic and ecological value, consequently causing increased pressure elsewhere. Wherever possible, steps should be taken to rehabilitate ecologically degraded natural features to restore the natural health of the areas concerned. The first step in the process of successful environmental planning is to recognise that development must be reconciled with the ability of the environment to sustain it.

Waste

The London and south east regional planning conference's *Revised Waste Planning Advice – Consultation Draft* (October 1996) proposes a bold but necessary solution to the increasingly difficult situation in the Region. Landfill capacity is likely to be exhausted within 10 years. It does not, however, deal with the increase in waste arisings likely to result from additional development predicted in the Region and this needs to be more fully addressed.

Water Quality

Water quality in the south east is already under pressure from current levels of development. Any increase in population in the south east would require high expenditure on infrastructure just to maintain river quality at its present level. In contrast equivalent increases in locations under less pressure would require less expenditure to maintain river quality. The Agency's three regions find it increasingly difficult to maintain existing standards of water quality let alone improve them. Further increases in population and development can only aggravate the situation through the effects of the capacity of sewage treatment works, urban runoff, eutrophication, dilution and recreation. Water quality is closely linked to water resources.

Air Quality

The additional development will require supplies of water and energy in the form of gas or electricity and waste disposal facilities.

There will be more demand for the products that are produced by industrial processes regulated under Integrated Pollution Control (IPC), for example power stations and waste incinerators. These processes have the potential to produce pollution, perhaps to several media at once which can be discerned over a wide area. Any such pollution will be added to that which arises from the activities of the inhabitants of these houses as they go about their business.

Therefore, it is important that when these IPC processes are developed their siting, design, construction, maintenance and operation are such that they do not create unacceptable burdens to an environment already under pressure. To help this the government has drafted a National Air Quality Strategy.

Minerals

Regarding the extraction of primary aggregates, the Agency is concerned about the growing impact of mineral workings on the water environment. In response, it has commissioned a series of environmental capacity studies of mineral bearing areas in the Thames Region, which suggests that the capacities of these areas are in danger of being breached in the medium to long term (i.e. 2006 and beyond) at current rates of extraction implied by mineral planning guidance (1994). Any increase in the required volume of material would accelerate this process if indigenous, land-won aggregate remains the principal source of supply.

9.6   The Measurement of Impact on the Environment

Thames Environment 21 is based on the assessment of the potential effect of each major growth location. Multi-attribute analysis is used to assess the impact of development proposals against principles of sustainable development. It enables the effects of development on a range of environmental criteria to be assessed thereby indicating the cumulative effect of development. As currently applied the technique is being used at a strategic level. It is likely to be developed for application in local environment agency plans to allow the joint analysis of both environmental and financial costs even when costs cannot be valued in monetary terms. The technique has the advantage of allowing the Agency to provide an indication of the environmental effects of development in a logical form for the benefit of local planning authorities.

Using a matrix the elements of each development location were assessed against the Agency's sustainability principles using the following notation: potential major beneficial effects, potential minor beneficial effects, potential minor negative effects, and potential major negative effects. Footnotes were also compiled to provide a qualitative element to the process.

The main findings and implications for the Agency were drawn for each of the major development location in a schedule entitled *Summary of Key Issues*. This highlighted the cumulative effects and linkages between development proposals or between sustainability principles.

9.7   Implementation through Partnership

The protection and enhancement of the environment relies on good working relationships between the Environment Agency, local planning authorities and the developers. The Agency is keen to work closely with them in promoting our plans and proposals. We believe that co-operation can offer significant benefits to the environment and to the community at large. Developers can likewise benefit from an enhanced environment. Such partnerships can therefore be an efficient and effective method for promoting the interests of all parties.

One of the primary pressures on the Agency will continue to be the impact of new development upon its interests. It is therefore essential for Agency staff to identify at the planning stages – be it pre-planning enquiries, planning applications or development plans – how the agency's interests can be protected and furthered through partnerships with external bodies including local authorities, industry and environmental groups.

The following examples show how environmental enhancements have been achieved through close working with local planning authorities and the development industry.

Soho Mill, Wooburn, Buckinghamshire

The River Wye is one of the most industrialised and urbanised within the western area of the Thames Region. Rising in the chalk downland to the west of High Wycombe, the river passes through the town centre before joining the River Thames at Bourne End. A number of mills were constructed on the river to provide power for local industry, and there have been many alterations to its course and nature as a result. Most of the mills ceased operating in the previous century, or adopted alternative sources of power. As heavy industry in the Wye valley declined the mill sites were redeveloped for alternative use or became derelict. As a result of the restrictive channels under buildings and around sites however, most mills place a restriction on the capacity of the river in times of flood.

Soho Mill is one such site, located in the village of Wooburn approximately three kilometres upstream of the confluence with the Thames. Following closure of the mill, the site was redeveloped for industrial use until this ceased in the 1980s and the buildings were demolished. The river through the site was left largely unaltered from its previous condition when the mill was working, and includes a two metre weir drop through the old mill race, together with restrictive brick channels and arches.

The developer wished to redevelop the main body of the site as a warehouse, and a new scheme of river works was designed. The developer was keen to enhance the river environment fronting his new premises, as were the local authority. A new design included splitting the weir into two smaller ones to lower a section of the river, and constructing a fish pass. Improvements to channel capacity were also included. The scheme is an excellent small scale example of the multidisciplinary Environment Agency work, and the benefits of working in partnership with the local authority through the planning process, and an enthusiastic developer.

Site of Former Chemical Works, Carshalton

This was the site of a former chemical works. The River Wandle used to run in a concrete channel from which abstractions and discharges to and from the site took place.

When the site became available for social housing the opportunity was taken to remove the concrete channel and culvert. The river has been restored near to its original course with a 'natural' bed and bunks and traditional footbridges have been provided.

Restoring a River as Part of a Major Residential Development at Aylesbury

The total site area is some 217ha and will be developed to provide a golf course, sports field, public open space and approximately 70ha of mainly residential development on the edge of the River Thames Floodplain.

In addition to a large flood compensation area to be excavated and landscape on the edge of the Thames floodplain, considerable work is being carried out to restore the heavily engineered main rivers to a more natural state. This involves reforming the watercourses as multi-stage channels within enhanced river corridors varying in width between 35 and 90 metres. The low-flow channels will be aligned with a natural sinuosity and provided with pools and riffles.

This development is an important milestone for the Region in representing the first major river restoration project achieved in co-operation with, and funded by, private development.

Millennium Site, Greenwich

In the year 2000, Greenwich hosted the Millennium Exhibition on the Greenwich peninsula site. Bounded by the River Thames to the north and north east, the highly contaminated and largely derelict site has a total of 2,200 metres of river frontage.

Million of visitors were expected to visit the Exhibition, which presents a unique opportunity for the Agency to demonstrate a best-practice model in urban flood defence works to maximise the recreational/landscape potential of the site whilst preserving and enhancing the conservation options. The Agency has identified the need to provide a robust, cost beneficial flood defence scheme but allow for an imaginative approach to the other needs of the site.

British Gas and now English Partnerships have been working closely with the Environment Agency to create a best practice riverbank scheme at the Millennium site to become renowned nationally and internationally. 1.24km of the existing river site frontage is known to be in a bad condition with an estimated life expectancy of less than 5 years and would need to be replaced as part of any redevelopment. The Environment Agency have been encouraging the developer to provide an innovative flood defence wall, incorporating some setting back to create enlarged beaches, an 'ecological sculpture', tidal terraces, timber fendering on vertical flood defence walls, beach replenishment/creation and improved habitats for potentially a multitude of wildlife.

A number of creative options have been identified and costed. As part of a riverside scheme, education signage, riverside paths and cycleways, and a permanent exhibition or 'water observatory' are also possibilities for the future to ensure that the riverside can be left as a permanent legacy for future generations to enjoy.

Blackwater Valley, Surrey and Hampshire

A major road development along the Blackwater Valley has been constructed during the late 1980s and early 1990s. This has necessitated the relocation of substantial lengths of river over a valley distance exceeding 10km. The existing channel was already substantially modified due to historical relocation as a result of floodplain gravel extraction, in some reaches the channel being more than five times the expected natural width. The road development provided an opportunity to recreate more natural characteristics. In 1995 a new length of channel was constructed as part of the overall road development near Aldershot. This was partially in virgin floodplain materials but involved filling a water-filled gravel pit and mounding the channel in new suitable fill material to the west of the new road. The recommended width and depth variables were recreated for the watercourse. Whilst the contractors were still on site, the bed of the river was raised at sites where riffles might be anticipated. This was achieved by placing several tonnes of angular coarse floodplain gravel on the river bed. The grain-size distribution was selected principally on the basis that it would remain in situ during moderate to high flows and would not be washed downstream. In general, this lowland river does not transport coarse bed load and natural riffles remain in situ.

Langley Court Nature Trails, Beckenham

The site was subject of a succession of planning applications for small developments, the cumulative effect of which was to increase the extent of impermeable cover of the site. None of the individual proposals provide adequate storage for surface water run off. One aspect of the proposals was the culverting of one branch of the brook. The developers agreed to provide an amenity lake for surface water storage, which has become an environmental asset associated with the new direction building. Wetland habitats have also been created. Upon completion of these works a nature trail has been completed.

9.8   Conclusion

In its role as a statutory consultee the Environment Agency relies on persuasion to secure enhancements to the environment. The scope for such improvements is limited once the developer has acquired his site. However, the scope is much greater if he is aware of the Environment Agency's requirements in advance through the development plan system. The Environment Agency has therefore developed a proactive approach to its planning work to take advantage of this situation.

In order to promote this proactive approach the former National Rivers Authority produced *Thames 21 – A planning Perspective and Sustainable Strategy for the Thames Region* which identified key catchment planning issues. Building on this work Thames Region has developed environmental assessment techniques which enable it to specify its requirements for site enhancement at the development stage. These techniques will also be used to enable the Agency to express a view on alternative locations for development which may be under consideration by local planning authorities when preparing their development plans.

The Agency's Thames Region has therefore strengthened its position for securing environmental enhancements through the planning system and the chapter has demonstrated a number of instances where this has been achieved.

Notes

1   The author wishes to thank the Environment Agency for permission to publish this paper. The views expressed are those of the author and are not necessarily those of the Agency.

2   Hugh Howes is a Chartered Town Planner and economist and has held the position of Principal Strategic Planner since 1991. Major responsibilities include articulating the interests of the Agency through Regional Planning Guidance and serving environmental enhancements through the development plan system. He is also Topic Leader for the Local Government Liaison Commission of the Agency's Research and Development Programme. Current projects include The Carrying Capacity of Catchments, Environmental Capital and Implementation and Planning Gain. He previously worked in the housebuilding industry and local government.

Chapter 10

Sustainable Management of the Physical Component of Urban Ecosystems: The Malaysian Perspective[1]

J.J. Pereira and I. Komoo[2]

10.1 Introduction

Since 1985, Malaysia has registered an unprecedented continuous economic growth of about 8 per cent per annum and this development has brought about numerous benefits including improved social amenities and a trend towards greater urbanisation of the population. The rapid economic development in Malaysia has contributed to environmental degradation and uncontrolled physical development especially in the urban areas. Protection of the environment is fast becoming a necessity rather than a luxury in order to maintain public health and well-being as well as to sustain the economic growth.

As in most developing countries, there are many challenges facing Malaysia's urban ecosystems, where the human, physico-chemical and biological environments are interlinked. One major challenge is unregulated physical development which has resulted in increasing occurrences of geohazards, causing property damage and high cost of maintenance as well as loss of lives, in extreme cases. This is a manifestation of poor planning and many of the geohazard problems in urban areas are often exacerbated by human activities.

This chapter focuses on sustainable management of the physical component of urban ecosystems, with emphasis on the geohazard prevention aspect, using the Klang Valley as an example. The overall aim is to highlight the socio-economic impacts of geohazards, the gaps in availability of geological information for planners and decision-makers, and to propose planning responses to minimise the effects of geohazards.

10.2 Issues and Challenges Facing the Klang Valley

The Klang Valley covers an area of about 2,100 km^2. It comprises part of 5 district councils (Gombak, Hulu Langat, Kuala Langat, Petaling and Sepang); 5 municipal councils (Petaling Jaya, Ampang Jaya, Klang, Shah Alam and Subang Jaya); and one city council (Dewan Bandaraya Kuala Lumpur). The whole valley is governed through

12 pieces of Federal legislation and some 10 sets of by-laws issued over the last 20 years. Local jurisdiction of the Klang Valley includes an almost linear zone of mainly urban and suburban centres, from Kuala Lumpur, through Petaling Jaya, Shah Alam, Klang and Port Klang where development is focused; with pockets of rural areas flanking this zone.

The environmental concerns in the Klang Valley are indicators of the types of problems that all major urban ecosystems in Malaysia face as a result of rapid development. The challenges facing the Klang Valley are related to the declining quality of the atmosphere and water as well as an increase in geohazards due to poor land management practices. Other challenges facing the Klang Valley include solid and toxic waste management, sewerage management, urban transport as well as socio-economic issues such as poverty, low cost-housing and squatter problems, among others.

Air Pollution

As a result of rapid industrial growth and the high demand for transportation, air pollution has reached unprecedented levels in the Klang Valley. This problem has been compounded by transboundary air pollution since the early 1990s. Some parameters of air quality in Petaling Jaya are shown in Table 10.1. The data show that in a five-year period the range and average pH values of rainwater have dropped slightly while the acidity has almost doubled. Heavy metals such as Fe, Ni, Cu, Zn and Hg in wet fallout from the air in Petaling Jaya have doubled in a five-year period (1990-1995), while there has been a ten-fold increase in the amount of Pb in the air. This data as well as other parameters monitored by the Meteorological Department of Malaysia, generally reflects the deterioration of air quality in Petaling Jaya and the Klang Valley.

Table 10.1   The pH and acidity of rainwater, and the heavy metal content of wet fallout in Petaling Jaya for 1990 and 1995

Parameter Measured (Unit)	1990	1995
pH Range	4.4-5.1	4.24-5.02
pH Average	4.6	4.5
Acidity (ueq/l)	6-43	24-76
Fe (ppm)	BDL*	<0.05
Ni (ppm)	<0.006	<0.017
Cu (ppm)	0.001-0.07	0.006-0.129
Zn (ppm)	<0.02	0.01-0.03
Hg (ppm)	<0.002	<0.003
Pb (ppm)	<0.004	<0.056

*BDL = Below Detection Limit

Source: Environmental Division, Meteorological Services Department of Malaysia (1995)

Water Pollution

The main source of water for consumption are rivers. In order to determine the Water Quality Index (WQI), five parameters are measured and they are: BOD, COD, ammoniacal nitrogen, suspended solids and hydrogen levels. Based on the WQI, the rivers are categorised as clean, slightly polluted or highly polluted. The Klang River is one of the 14 highly polluted rivers in Malaysia (DOE, 1996). The main causes of water pollution are domestic sewage, industry, highlands development and land clearing activities. Highlands development and land clearing activities have resulted in an increase in suspended solid and changes in river morphology, and have also contributed to increased flooding and pollution of coastal zones.

Geohazards and Land Management

The Klang Valley urban ecosystem has rapidly expanded in the past decade due to the economic growth. As most of the suitable building land has been utilised, new developments are increasingly located in less suitable sites, such as unstable slopes, floodplains and over limestone bedrock or ex-mining areas. These areas are prone to geohazards such as landslides, flooding and subsidence. This has inevitably led to rapidly escalating costs due to damage, provision of remedial measures and rebuilding, to both the government and society.

There has been growing concern over land management practices in the Klang Valley due to housing and industrial development. Land clearing activities have exposed larger areas susceptible to erosion, resulting in increased sedimentation of rivers and subsequent flash-flood in the Klang Valley. The increase in the volume of surface water run-off due to land clearing has also contributed to flash-floods and landslides. State governments and local councils are constantly being urged to apply environmentally sound land management practices to reduce these problems in the long-term.

Other Concerns

Solid waste disposal, traditionally carried out by local councils, has recently been privatised. The aim of this privatisation is to incorporate recycling and other elements of environmental consideration into one integrated solid waste management system (Phang and Koh, 1996).

Major sources of industrial pollution in Malaysia are the metal finishing, electrical and electronics, textiles, food processing, chemical, palm oil, rubber, wood as well as iron and steel manufacturing industries (RM7, 1996). These industries are concentrated in the Klang Valley and as a result there has been an increase in the use of toxic chemicals and generation of toxic waste. In order to arrest this problem, the management of toxic chemicals and hazardous waste is being strengthened. The establishment of a centralised and integrated facility for storage, treatment and disposal of hazardous waste serves to improve the management of toxic waste.

The Klang Valley generally has a high level of standard basic sewerage facilities. The sewerage management has been privatised in order to improve its efficiency, provide uniform functions and extend its coverage beyond the boundaries of local councils. However, government supervision of this privatisation scheme is still required in order to improve urban management so that a healthy environment can be achieved.

The traffic situation in the Klang Valley is acute. In 1995 the number of vehicles plying the roads on a working day was estimated at 1.4 million (RM7, 1996). A number of road projects and the construction of an integrated public transport system have been undertaken to improve traffic flow in the Klang Valley.

10.3   Geohazards in the Klang Valley

Geohazard has been defined as 'a hazard of geological, hydrological or geomorphological nature which poses a threat to man and his activities' (Doornkamp, 1989). Geohazard has also been defined as being 'one that involves the interaction of man and any natural process of the Planet' (McCall, 1992). Based on the second definition, all processes that involve the solid Earth, oceans and atmosphere, including geophysics, meteorology, climatology and glaciology, come under the scope of geohazards.

Geohazards include volcanic eruptions, earthquakes, tsunamis, floods, coastal and fluvial erosion, landslides, ground subsidence and dam failures. Geohazards can be either natural or human-induced. Currently, natural geohazards are the major cause of loss of life and damage to property. However, human-induced geohazards are increasing rapidly and may eventually outnumber natural geohazards as the main cause of loss of life and damage to property (McCall, 1992). The difference separating a natural and human-induced geohazard is sometimes indiscernible, especially in conditions where a catastrophic event is made worse by the intervention of man.

The factors that have increased the occurrences of geohazards can be divided into two categories. The first category of factors is related to the purely materialistic pursuit of development which result in excessive emphasis on commercial development, increased technological development and industrialisation and increased scientific tinkering with Nature without concern for possible long-term effects or disaster. The second category of factors is related to management issues and include increased population density, poor engineering practices and insufficient or poor prior geological investigation in construction, water management and waste disposal (McCall, 1992).

Geohazards Occurrences

Geohazards have been studied in the Klang Valley since the late 1970s, under the topic of urban geology; the study of geological and geo-technical problems in urban areas. The emphasis then was on reporting the results of technical investigations after the geohazard had occurred (Shu and Chow, 1979; Shu, 1986) and highlighting the importance of engineering geological studies for urban development (Tan et al., 1980 and Tan, 1986).

Geohazards that occur in the Klang Valley are landslides, ground subsidence, and fluvial hazards, which include both floods and river bank erosion (Pereira et al., 1996; Komoo et al., 1996). The locations of these geohazards are illustrated in Figure 10.1.

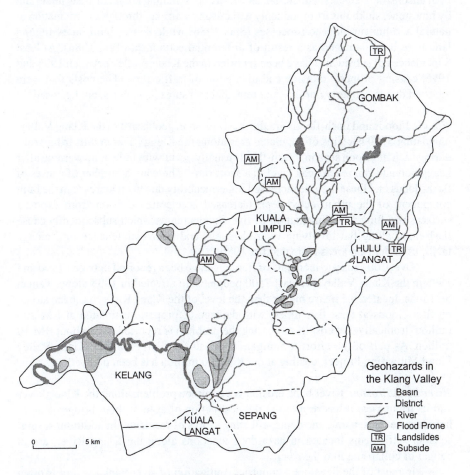

Figure 10.1   Geohazards in the Klang Valley
Source: After Ibrahim et al. (1996)

Landslides The combination of heavy rainfall and removal of vegetation results in landslides in the hilly areas of the Klang Valley. Landslides have occurred in at least 20 localities, varying in the speed, depth and content of moving material. The total cost of damage to property in the Klang Valley due to landslides is not known. The cost of slope stabilisation at *one locality* (Taman Cuepacs, Cheras) was estimated at

RM 12 million. The most catastrophic landslide-related incident was the collapse of the Highland Towers in Ampang where 48 people perished.

Subsidence Subsidence is the sudden or gradual sinking of the ground surface. In the Klang Valley, subsidence is known to occur in areas underlain by limestone, in areas with unconsolidated clayey material, as well as in ex-mining areas. In areas underlain by limestone, sinkholes form suddenly without any warning, through a combination of natural and human-induced processes (Shu, 1986) while houses built on ex-mining land have been damaged as a result of differential settlement (Tan, 1986). At least 8 incidences of subsidence have been reported in the Klang Valley between 1993 and 1995, causing property damage and disruption of traffic flow. The costs that were incurred due to property damage and remedial measures have not been disclosed.

Floods Floods and flash-floods are the most common geohazard in the Klang Valley, inundating 5700 hectares of populated area along the Klang River and its tributaries, during each monsoon season. Flash-floods usually occur with little or no warning after heavy rain and can reach peak level in a short time. There are a number of causes of flash-floods and these include increased impermeability due to an increase in the built component of the urban ecosystem; increased accelerated erosion from exposed surfaces resulting in sedimentation of rivers and streams which subsequently cause flash-flood along flat low-lying river channels; and poor maintenance of drainage facilities in built-up areas (Jamaluddin, 1986).

Over 100 separate incidents of flooding have been reported between 1990 and 1995 in the Klang Valley. At least 7500 people were affected in 1995 alone. Taman Sri Muda, located 0.5 metre higher than the level of the Klang River, has been struck by flash-floods at least five times, with damage of property estimated at RM 200 million. Ironically, the cost of mitigating the problem is reported to be about RM 10 million. As part of the effort to mitigate the flash-flood problem, the Klang Valley Flood Mitigation Project, costing about RM 760 million has been planned.

River Bank Erosion River bank erosion is a common problem along the Klang River and its tributaries. It has resulted in the destruction of at least three bridges and 14 houses in seven separate incidents, affecting scores of residents. In addition, several squatter settlements located in hazardous locations along the Klang River, are in danger of collapsing into the river.

In most of the developed countries, mitigation of river bank erosion is often combined with flood mitigation projects. In the US, the approaches practised include building of engineering structures, control of river flow, creation of greenways along the river, controlling wetlands development and controlling occupation of floodplains. Building of engineering structures to control river bank erosion has become increasingly unpopular in recent times as it does not provide a long term solution and instead transfers the problem to another location (Gares et al., 1994). As engineering efforts to stabilise the Klang River and its tributaries are limited, it is anticipated that the people who occupy the river banks, mainly squatters, will have to absorb the losses and relocate.

10.4   Geohazards Management Approaches

It has been proven world-wide that safe development can be carried out in areas with geological hazards if geological information is used in project planning (Brook, 1992 and Tyler, 1994). Ignoring geological processes and omitting them from planning considerations result in unsustainable development and can be expensive in human and economic terms.

The planning responses to natural and human-induced geohazards in Peninsular Malaysia are not as advanced as in countries such as the USA, UK or Italy (Cogeoenvironment, 1995; USGS, 1996; DOE-UK, 1990). The local councils in the Klang Valley take minimal account of the safety element in planning. The management practices need to be improved in order to achieve sustainable human development. Legislative provisions for improving this situation and other shortcomings have lately been enacted.

Current Approaches

In 1976 structure and local plans were introduced in Malaysia under the Town and Country Planning Act, 1976 (Act 172). The Kuala Lumpur Structure Plan (KLSP) was implemented in 1984 in accordance with the Federal Territory (Planning) Act, 1982 (Act 267). The KLSP includes a statement of principles and public policies, strategies for development, land-use, and social, economic and physical improvements as well as traffic management for 20 years, up to the year 2000. Structure plans for the other parts of the Klang Valley contain principles, policies and strategies in accordance with the development aspirations of individual districts/municipals, as determined by the State. Local plans have been drawn up for several areas in the Klang Valley, but the majority are still in the process of being formulated.

Although structure plans and local plans are largely based on physical parameters, geological and geomorphological information pertaining to potential geohazards are not taken into account when these plans are drawn up. This has led to increasing occurrences of geohazards, especially in rapidly expanding urban centres in the Klang Valley. For example, the housing estate of *Taman Sri Muda* is located on a floodplain. Floods in this area have caused about RM 200 million in losses. If incorporation of geological information was part of the planning process, the question of whether this site is suitable for a housing estate would have risen. At the very least, proper mitigating measures could have been installed to avoid economic losses.

Among the key issues necessary for improving the planning process to achieve sustainable human development include socio-economic considerations, management of resources, and suitability of land taking into account its potential geohazards and environmental impacts. Strategies and plans for improved planning of the environment and management of the resources as an integral part of the socio-economic development plans have to be formulated. In the case of Malaysia, ignorance of geological processes and their omitment from planning considerations have resulted in unsustainable development. It has been an expensive lesson in human and economic terms (for instance the Highland Towers tragedy, Taman Sri Muda floods).

There is an urgent need for geohazard risk identification in the planning process to ensure a safe environment for the well being of present and future generations. It should be taken into account and integrated into the overall planning and development process, and the capacity of all planning and executing agencies at the district and local levels should be strengthened in this respect.

Legislative Measures

The Town and Country Planning Act, 1976 (Act 172) has recently been amended with the objectives of strengthening the administrative mechanism of planning at the State and local council levels in order to improve efficiency and meet the current changing needs for planning; balancing environmental conservation with physical development; and reducing damage to property and loss of life due to uncontrolled development. The amendment is referred to as the Town and Country Planning (Amendment) Act 1995 (Act A933); has been accepted by all the States apart from Kelantan (Ibrahim, 1996), and shall come into force on such date as may be appointed by the State Authority, by notification in the State *Gazette*.

One of the new sections of the Town and Country Planning (Amendment) Act 1995 (Act A933), issued as Section 21A requires the submission of a development proposal report, when applying for planning permission under Section 21(1). The development proposal report should contain description of the land including its physical environment, topography, landscape, geology, contours, drainage, water bodies and catchment as well as natural features of the land, among other requirements. The applicant has to show to what extent the development proposed would affect the above elements and adjacent areas. The applicant is also required to suggest measures to conserve and improve the physical environment. The requirement of a development proposal report increases the power of the local councils and enables the evaluation of suitability of the proposed project in terms of planning, for conservation and physical development, by incorporating elements of safety, health and well-being (Ismail, 1996).

The Town and Country Planning (Amendment) Act 1995 (Act A933) is the first line of defence in terms of environmental protection, and it can also be utilised as a planning response to geohazards. The provision of appropriate and easily understood geological information regarding potential geohazard occurrences in the development proposal report, will aid the local councils in making a decision regarding the application for planning permission.

Availability of Information for Planning

Information about the physical aspects of geohazards aimed at educating non-geologists in the policy arena about risk assessment and disaster reduction is scarce in Malaysia. In general, the managers, planners and decision makers have yet to fully understand and appreciate the need for adequate geological input in the siting of a project, in order to prevent or reduce deaths, injuries and property damage.

The Geological Survey Department (GSD) provides geological information for land-use planning at the district and local levels and for specific development projects of national interests. The role of the Geological Survey Department in providing geological information on geohazards for the preparation of structure and local plans, is overshadowed by its role in providing information on the availability of mineral resources. The provision of geological information on geohazards at this level should be further enhanced, to attain the standards set when providing information for specific development projects. The GSD will publish terrain classification maps for selected areas in the Klang Valley at a scale of 1:10000. This may prove to be very useful for planning purposes.

Geohazard maps have been constructed for Pulau Pinang and Seberang Perai, as part of a study in 1992-1994, commissioned by the State government and executed by Universiti Sains Malaysia (Tjia, 1995). The maps identify potential geohazards such as fractured rocks in fault zones, ground subsidence in peat, landslide and rock fall prone areas. This commendable effort should be seriously examined by the Town and Country Planning Department and local councils for planning considerations, especially in the formulation and revision of structure and local plans.

The Institute for Environment and Development (LESTARI) has commenced research on 'An Integrated Approach to Urban Ecosystem Management' in order to address in a holistic manner, the management of emerging social and physical problems facing major urban centres. One major component of this project is to identify the extent of various potential geohazards in the Klang and Langat valleys; distinguish mechanisms, processes and factors contributing to the geohazards phenomena; and recommend approaches, policies and guidelines for the management of geohazards.

The outputs expected include a comprehensive database on geohazard occurrences (which is currently not available in Malaysia), geohazard maps and associated reports for the Klang Valley. The Geological Survey of Malaysia has agreed in principle, to collaborate on this aspect of the research. In the long term, it is hoped that this project will increase awareness among the planners, managers, decision-makers and society of the essential contribution of geological understanding to sound planning and management of geohazards, especially in urban areas.

10.5 Conclusions

In order to achieve sustainable human development, it is necessary to ensure that the planning process includes socio-economic considerations, management of resources, and suitability of land taking into account its potential geohazards and environmental impacts. Strategies and plans for improved planning of the environment and management of the resources as an integral part of the socio-economic development plans have to be formulated.

The provision of adequate geological information is vital in the planning stage or siting of a project so that planning responses can be taken to minimise the effects of geohazards. This will serve to prevent or reduce deaths, injuries and property damage which costs the nation millions of *ringgit* annually. On a project scale, geological information can be provided in the development proposal plan that is required under the Town and Country Planning (Amendment) Act 1995 (Act A933), when applying for planning permission.

An effort should also be made to help planners and decision-makers to understand and appreciate risk assessment and disaster reduction due to geohazards. Creating effective linkages between geologists, planners and decision-makers will increase the awareness of the essential contribution of geological understanding to sound planning in order to reduce or mitigate geohazards. Geohazard risk assessment should be incorporated as part of the planning process, when drawing up structure and local plans.

The research project on 'An Integrated Approach to Urban Ecosystem Management' seeks to address problems in the Klang Valley in a holistic manner. The geohazard component of this project, carried out by LESTARI in collaboration with the Geological Survey Department of Malaysia, will see the compilation of an inventory of geohazards and related information for modelling of risks associated with geohazards. This data will then be integrated with social and economic imperatives. Lessons learnt from the Klang Valley will be applied in the development of the Langat Valley, where future development activities will be focused, to complement the Putra Jaya (new administration capital of Malaysia), new international airport and multi-media super corridor projects.

Notes

1   The research on 'An Integrated Approach to Urban Ecosystem Management' is funded by IRPA (08-02-02-0002), under the Urban Ecosystem Management Programme, Institute for Environment and Development (LESTARI), Universiti Kebangsaan Malaysia.

2   Joy J. Pereira is lecturer at the Institute for Environment and Development (LESTARI), Universiti Kebangsaan Malaysia and her research focuses on planning responses for geohazard prevention and environmental management in the mining and construction industries. Ibrahim Komoo is Professor and Associate Director at the Institute for Environment and Development, Universiti Kebangsaan Malaysia where he leads the Urban Ecosystem Management Programme. His current research projects are on integrated planning of urban ecosystems, conservation geology and geohazard prevention.

References

Brook, D. (1992) Policy in Response to Geohazards: Lessons from the Developed World?, in G.J.H. McCall, D.J.C. Laming and S.C. Scott (eds) *Geohazards-Natural and Man-Made*, London: Chapman & Hall, pp. 197-208.

Cogeoenvironment (1995) *Planning and Managing the Human Environment – The Essential Role of the Geosciences*, Commission on Geological Sciences for Environmental Planning, IUGS, Haarlem, The Netherlands.

DOE (1996) *Malaysia Environmental Quality Report 1995*, Department of Environment, Ministry of Science, Technology and the Environment, Kuala Lumpur.

DOE-UK (1990) Planning Policy Guidance: Development on Unstable Land, *Department of the Environment/Welsh Office Planning Policy Guidance Note (PPG) 14*, HMSO, London.

Doornkamp, J.C. (1989) Hazards, in G.J.H. McCall and B.R. Marker (eds) *Earth Science Mapping for Planning, Development and Conservation*, Graham and Trotman, London, pp. 157-173.

Gares, P.A., D.J. Sherman and I. Nordstrom (1994) Geomorphology and Natural Hazards, in M. Morisawa (ed.) *Geomorphology and Natural Hazards*, Elsevier, Amsterdam, pp. 1-18.

Komoo, I., J.J. Pereira and S. Maziah (1996) Geobencana di Lembangan Klang. *Prosiding Seminar Geologi Sekitaran, Bangi, 7-8 Dis. 1996*. Organisers: Persatuan Geologi Malaysia dan Jabatan Geologi, UKM, pp. 74-84.

Ibrahim, I. (1996) Pindaan Akta Perancangan Bandar dan Desa dan Laporan Cadangan Pemajuan (LCP). *11th Senior Officials Meeting of the Town and Country Planning Department, Kuching, Sarawak, 2-5 October, 1996*. Organisers: Town and Country Planning Department of Peninsular Malaysia and the Government of Sarawak.

Jamaluddin, M.J. (1986) Human Perception and Responses to the Flash-Flood Hazards in Parts of Kuala Lumpur, in S. Sani (ed.) *A Study of the Urban Ecosystem of the Kelang Valley Region, Malaysia, Vol. 1 – The Kelang Valley Region: Some Selected Issues*, Universiti Kebangsaan Malaysia, Bangi, pp. 84-104.

McCall, G.J.H. (1992) Natural and Man-Made Hazards: Their Increasing Importance in the End-20th Century World' in G.J.H. McCall, D.J.C. Laming and S.C. Scott (eds) *Geohazards-Natural and Man-Made*, Chapman & Hall, London, pp. 1-4.

Pereira, J.J., I. Komoo, and M. Sualaiman (1996) Geohazards and the Urban Ecosystem. *Proceedings-Forum on Geohazards: Landslide & Subsidence, Universiti Malaya, 22 October 1996*. Organiser: Working Group on Engineering Geology and Hydrogeology, Geological Society of Malaysia. pp. 3.1-3.13.

Phang, S.N., and A. Koh (1996) Urban Environmental Management in Malaysia. *Workshop on Urban Environmental Management, Jakarta, Indonesia, 19-23 August 1996*. Organisers: German Technical Cooperation Agency (GTZ) and ASEAN Senior Officials on Environment (ASOEN).

RM7 (1996) *Seventh Malaysia Plan 1996-2000*, Economic Planning Unit, Prime Minister's Department, Kuala Lumpur.

Shu, Y.K. (1986) Investigation of Land Subsidence and Sinkhole Occurrences in the Klang Valley and Kinta Valley, Peninsular Malaysia, in B.K. Tan and J.L. Rau (eds) *Landplan II Proceedings-Role of Geology in Planning and Development of Urban Centres in Southeast Asia*, Asian Institute of Technology, Bangkok, pp. 15-28.

Shu, Y.K., and W.S. Chow (1979) Land Subsidence at Serdang Lama, Selangor. *Unpublished Report E(F)4/1979, Geological Survey of Malaysia*.

Tan, B.K. (1986) Geological and Geotechnical Problems of Urban Centres in Malaysia, in B.K. Tan and J.L. Rau (eds) *Landplan II Proceedings-Role of Geology in Planning and Development of Urban Centres in Southeast Asia*, Asian Institute of Technology, Bangkok, pp. 10-14.

Tan, B.K., and J.K. Raj (1980) Geotechnical Engineering Development in Malaysia. *Proc. Malaysia Scientific Congress*, Congress of Malaysian Science and Technology, Kuala Lumpur, pp. 63-74.

Tjia, H.D. (1995) Terrain and Geohazard Maps of Negeri Pulau Pinang and Seberang Perai. *Proceedings-Forum on Environmental Geology and Geotechnics, Universiti Malaya, 24 October 1995*. Organiser: Working Group on Engineering Geology and Hydrogeology, Geological Society of Malaysia.

Tyler, M.B. (1994) Look before You Build, *US Geological Survey Circular 1130*.
USGS (1996) US Geological Survey Natural Hazards Programs: Lessons Learned for Reducing
    Risk, http://h2o.usgs.gov/public/wid/html/HRDS.html.

Part C
City-wide Approaches to Integration

Chapter 11

Environmental Protection in the Free State of Saxony and the 'Ecological City' Model Project

A. Vaatz[1]

11.1   Introduction

The approach of combining environmental protection issues with regional planning tasks is still rare among the governments of the German federal states. Yet it has proven its value especially during the reconstruction of the Free State of Saxony, one of the five new federal states in the east of Germany. This has given us the opportunity to immediately combine ecological and economic concerns during the development of the infrastructure and the economy, which has also enabled us to consider many possibly conflicting goals.

This is the seventh year of the reunification of Germany. We have just completed one of the most radical economic changes in history as well as tackling the extraordinarily difficult task of rectifying forty years of ecological destruction wrought by the socialist planned economy. The level of environmental protection that has been put in place during the first six years since reunification is such an extraordinary achievement that no one would have seriously considered it previously. In 1990 we were confronted with an utterly hopeless situation that we had inherited from the former socialist state.

More than 60 per cent of the lakes and rivers of the Free State of Saxony had to be classified as seriously or severely polluted! Only one in every five communities possessed a sewage plant and most of these were in a catastrophic condition, discharging mainly untreated water into the receiving waters. Approximately 25 per cent of the population were drawing their drinking water from public sources which were supplying drinking water of an unsatisfactory quality. There were about 1,800 waste dumps on the relatively small territory of the Free State of Saxony and these contained domestic and industrial waste without any measures in place to protect the soil and the ground water. At a total of 2 megatons, the sulphur dioxide emissions in Saxony were twice as high in 1989 as the total amount in all of West Germany.

11.2 Some Major Recent Environmental Accomplishments

Today, only seven years later, things have definitively changed for the better. Here are just a few examples.

More than €1.25 billion have been invested since 1994 in the supply of drinking water. We are now in a position to comply to a substantial extent with the stringent health requirements of the Drinking Water Ordinance. Almost 300 sewage plants have been either newly built or completely reconstructed since 1990. This has brought about a marked improvement in the state of the lakes and rivers, with the share of seriously polluted inland waters being reduced by half. Of the former 1,800 waste dumps, only 50 dumps remain today for domestic waste. These have been refurbished in the previous decade and equipped with the necessary facilities. The former 88 coal-fired power plants in Saxony were all without desulphurisation systems and dust control equipment. Some were located in the middle of densely populated areas. Strict emission controls and the forced closure of half of the plants have resulted in noticeable improvements.

In Saxony at the turn of the century, we were able to achieve emission rates for sulphur dioxide and dust which will be up to 95 per cent less than the rates measured in 1989. This enormous success has its price of course. In 1994 alone the public sector, the economy and private consumers had to pay almost €5 billion for measures designed to improve the environment. That is more than €1,000 per head of population.

The Saxon environmental policy after German reunification was characterised by damage limitation. An antiquated industrial structure and the use of brown coal containing sulphur in densely populated and highly industrialised areas of the Free State of Saxony were the main causes of the catastrophic amount of air pollution. Since the worst affected areas were easily identifiable and solutions were already at hand, we were able to put successful measures in place rather quickly.

Over the next few years we will concentrate on sustaining the results achieved and planning for the future. One of the most pressing issues in this respect is further development in the urban areas as the cities harbour the greatest potential for saving energy, both in the use of energy as well as in the total surface requirements. The urban areas thus offer many possibilities for a more ecological use of land and the establish-ment of more efficient urban transport facilities as well as a more efficient use of energy and other resources. We must further aim at reducing the amount of waste arising and water used in the densely populated urban centres as well as continuing to strive to use more ecological building methods and to improve the quality of the city air.

11.3 The Effects of Increasing Road Traffic

The quality of the city air in particular is a major issue where the success of recent years is endangered by an ever-increasing volume of traffic. The number of motor vehicles registered in the Free State of Saxony has almost doubled since 1989.

Passenger traffic and also especially the volume of freight traffic on the roads will dramatically increase over the coming years. In 2005 private use of cars will have doubled compared with 1989 and there will be four times as much freight traffic on the roads. At the same time, by the year 2005 the number of rail passengers will have sunk to only half the number of 1989 and the amount of freight traffic on rail will have decreased even more dramatically, by four fifths.

This exceptional development will mainly have an impact on the cities. Yet one must not forget that for over 40 years there has been no sensible traffic policy in place in the cities of East Germany. What they do possess are relatively well-developed public transport systems which we are maintaining and trying to expand according to our means.

1990 was also the year in which our dilapidated roads were hit by an avalanche of cars which we have been unable to cope with up to this very day and which we will not be able to withstand in the way of our 'car-friendly cities' of the 1960s. This is why it is essential to develop ecologically sensible traffic programmes that will be able to meet the predicted future traffic levels without neglecting the demands of present-day environmental protection policies. Even though the former East German car pool has now been overhauled with the 'Trabbi' being gradually withdrawn from public life, the pollution effects attributable to the increasing amount of motor vehicle traffic are still expected to increase.

The occurrence of some pollutants is expected to diminish up to the turn of the century, whereby the levels of carbon monoxide and hydrocarbons in particular are among those due to decrease over the years. Yet we still expect nitrogen oxides to retain their high levels of emission. As you are already aware, these are essentially the results of the change-over in East German cars from two-stroke to four-stroke engines and the increase in the number of catalytic converters for cars. Therefore, we cannot expect a decrease in the levels of ozone during the summer in the medium term and the increasing acidification of the soil will also remain a problem.

Road traffic also contributes to a large degree to the consumption of energy and thus to the CO_2 emissions. CO_2 emissions from road traffic have increased by 70 per cent since 1989 and are expected to rise by another 10 per cent up to the year 2005. The share of road traffic in the total energy consumption has dramatically increased in the last years and now stands at approximately 30 per cent, a level expected to continue to rise. Further consequences of the tremendous increase in road traffic is the noise problem, with road traffic now being the number one source of noise in Germany.

The issue 'road traffic and the environment' is currently at an impasse in Germany. Neither a policy of a standard speed limit on the motorways nor a policy of markedly increasing the tax rate applied to petrol have been able to gain a necessary political majority. Progress will probably only be made by converting the tax-deductible mileage into a mileage concession applicable to other modes of transport. To get some movement back into the deadlocked situation, we want to work on the area with the greatest need and where the pressure from the population is also the greatest: urban development.

11.4　The 'Ecological City' Project

This is the reason why the Saxon Ministry for the Environment and Regional Development has launched an integrated pilot project designed to demonstrate the possibilities of a long-term and sustainable urban development scheme. The pilot project 'Ecological City' has been designed to take a model community to serve as an example for as many other similar cities in Saxony as possible. We would like to select a community for this pilot project whose initiatives and ideas we can then assist with advice and funds. Priority will be given to measures leading to a reduction in traffic-related pollution, conservation of natural resources and natural climate.

The goal is to be able to carry out an integrated municipal traffic scheme with measures aimed at planning, directing and limiting the flow of traffic. In particular the private use of cars for leisure purposes is to be reduced and the necessary commercial traffic, such as that of trades and business people, is to be organised more efficiently. Motorisation technology with especially low emission factors and efficient uses of energy are to be promoted and environmentally-friendly means of transportation, such as public transport and also bicycles and pedestrian traffic, are to be increased as far as possible.

A whole complex of meaningful policies exist, not all of which I can mention at this point – you are all familiar with the wide range of modern urban development concepts. Simulation models, for example for air and noise emissions, will help us identify and evaluate individual control measures.

Main issues will include co-ordinated urban development planning enabling a diversification of domestic housing and commercial buildings to address the commuting problem and to conserve energy. Energy conservation measures will include the use of highly efficient modern long-distance energy supply systems combining heat and power as well as the improvement of insulation for buildings. This is especially important in view of the numerous old buildings still in use in East Germany. At the same time, the latent potential of regenerative energy must be harnessed.

We finished the selection procedure in 1996. Few of you will be familiar with the Free State of Saxony and its marvellous cities. I am very tempted therefore to avail of this opportunity to present you with a comprehensive introduction to Saxon history, culture and geography. What a pity therefore that I will only be able to cover a few major highlights which I feel are most important about the cities in our model project.

We have enabled all Saxon cities whose populations exceed 5,000 inhabitants to participate in the selection process. With a response rate of 30 per cent and a total of 41 applicants, we feel that we have had a very good response. Following several selection rounds and discussions with the communities' representatives regarding the completion of their individual projects, we were able to short-list three cities: Görlitz, situated near to the Neiße River and bordering on Poland; Lichtenstein in close proximity to Zwickau; and Taucha, situated to the north-east of Leipzig.

After considering all aspects, Taucha was finally selected as our candidate. Our main concern was that the project tasks remain simple and clear and therefore transferable to as many other Saxon cities as possible. We also carefully took the city management's commitment and the city council's backing into consideration.

We were convinced of Taucha's sincerity by its previous activities in environmental issues and its sound and integrated total concept. This aspect of the amount of preparation which has gone into the project and the initiative of the community is very important to us as we do not want to implement a set strategy from above in this pilot project but rather support the cities' own initiative. Taucha will be focusing on the areas of clean air, noise reduction, traffic control and urban development as well as improving the city's living conditions and its recreational facilities. The city of Taucha has been especially hard hit by the extreme burdens imposed by several large-scale industrial projects carried out in the metropolitan Leipzig area. Taucha also lies in the flight path of the international Leipzig airport and suffers greatly from the amount of traffic passing through on a particularly heavily used federal highway. The city can especially develop in terms of domestic housing and commercial buildings spilling over from Leipzig and on account of its fallow land in the city centre. With its population of 14,000 inhabitants, the tasks confronting Taucha are clear enough and can be easily applied to other cities. Saxony has 36 cities with a population of between 10,000 and 20,000 inhabitants and a further 15 cities with up to 30,000 inhabitants. Görlitz also convinced us by its thorough preparation work and it also has measures in place containing excellent ideas for co-operating with the neighbouring city on the Polish border. The decisive factor in our final decision was the size of Görlitz with its approximately 67,000 inhabitants, a size which we did not consider to be so suitable for our demonstration purposes.

Finally Lichtenstein impressed us with its integrated concept for improving the city's air quality, its scheme for improving the quality of the city's streams and ponds and for protecting the biotope as well as its ideas for additional landscaping measures. But since Lichtenstein is primarily focusing on landscape protection, this to a certain extent limits the way in which these activities could be applied to other cities in Saxony.

We will discuss with representatives from the city of Taucha further measures for implementing the pilot project. My ministry finds the following points of special significance: a noise reduction plan, reporting on all local sources of air pollution, a traffic scheme which addresses the issues of reducing noise and pollutants, changing public building heating from coal to natural gas, and allocating sites for wind-power stations in the land-use map.

11.5   Conclusion

The results and experience gained from the pilot project 'Ecological City' will be published at regular intervals of course. We hope they will therefore animate others to carry out similar schemes. The project is intended to run for several years. I would be pleased if our 'Ecological City' Taucha secures international recognition. If one leaves out the East German peculiarities, there are definitely quite a few points which can be transferred to other cities of a similar size confronted with similar problems.

Note

1   A. Vaatz is State Minister for the Environment and Regional Development of Sacksen (*Sächsisches Staatsministerium für Umwelt und Landesentwicklung*), Germany.

Chapter 12

Lessons from an Adaptation of the Dutch Model for Integrated Environmental Zoning (IEZ) in Brooklyn, NYC

H. Blanco[1]

12.1 Introduction

This chapter presents the planning implications of the Baseline Aggregate Environmental Loads (BAEL) study that adapted and applied the Integrated Environmental Zoning (IEZ) method, which the Dutch Ministry of Housing, Physical Planning and the Environment (VROM) developed in the early 1990s, to a community in New York City, Greenpoint-Williamsburg. The chapter provides a brief community profile of Greenpoint-Williamsburg, followed by an explanation of IEZ and how it was modified in the study. It then goes on to present some of the major applications of this approach, including in siting, zoning, land use, resource allocation, and environmental regulatory processes. It concludes with a comparison between IEZ and BAEL, and suggests their implications for planning.

12.2 The Community

Greenpoint-Williamsburg is an old industrial/working class area in northern Brooklyn, close to five square miles, which houses over 155,000 people, provides over 35,000 jobs, and has one of the highest concentrations of environmental burdens in New York City.

The District's population resides in distinct ethnic communities, which were brought together as Community District 1 in Brooklyn in 1967, when the NYC Department of City Planning created a city-wide system of community planning districts. Greenpoint's population has a concentration of Poles and other Eastern Europeans; the Southside has a concentration of Latinos, especially Puerto Ricans; the Northside is a more mixed community with a recent influx of artists attracted by affordable loft buildings.

South Williamsburg has a predominance of Hasidic Jews. Overall, the US Census reported that, in 1990, 46 per cent of the district's population was White, 44 per cent Hispanic, 7 per cent African American and 3 per cent Asian and other.

The residential population is primarily working class or poor. This is reflected in income and educational attainment. In 1990, the median income was $11,000 less than New York City's median income, and 36 per cent of the district's residents were below the poverty line. Educational achievement is also below the city profile with lower percentages of people age 25 and over completing high school or college. When compared to the city's profile, a significantly larger percentage of the district's population is employed in manufacturing or other blue-collar occupations.

Industrial and transportation and utilities uses, uses with significant environmental impacts, make up approximately 60 per cent of the net area of the district. These uses are primarily found encircling the three-sided waterfront. Although jobs in the industrial sector have been declining over the last few decades, the area remains one of the two major manufacturing centers in the City. A 1993 NYC Planning survey estimated that the area generated over 34,000 jobs in its industrial zones. See Figure 12.1 for the existing land use map. Residential uses, mostly in one to three unit frame buildings and small tenements of three to five stories in height, are concentrated in the central areas of the district, although there are many areas with a fine-grained mix of industrial and residential uses. Due to the larger household size and the relatively small size of the housing stock in the District, overcrowding is a problem.

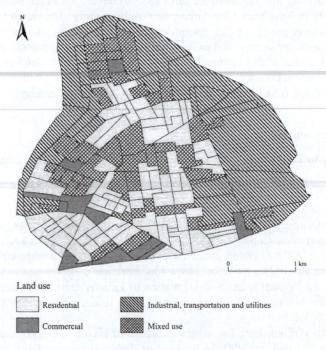

Land use

▢ Residential   ▨ Industrial, transportation and utilities

▮ Commercial   ▨ Mixed use

Figure 12.1   Land use map of Greenpoint-Williamsburg
Source: RPAD (7/87), DCP-field-work (10/88), and Hunter Studio-field-work (3/95)

Greenpoint-Williamsburg has a history of industrial production and resulting pollution dating back to the early 1800s, when the area's water-front first emerged as a strong shipping, industrial and transportation center for the City. Industries in the area included oil refineries, metal fabrication, and shipbuilding. Over the past few decades, as industrial activities have declined throughout the country due to economic restructuring, industrial uses have declined in the district as well. The rail lines that provided direct access to the waterfront piers have been dismantled, many of the piers have been abandoned, and factories have left, leaving abandoned buildings and vacant lots along the waterfront.

Today, the district is burdened with a formidable set of environmental problems: the largest underground oil spill in the history of the United States (larger than the Valdez spill) discovered in the late 1970s, which is a Superfund site; the polluted waterways of Newtown Creek and English Kills with high concentrations of metals and PCBs; the largest sewage treatment plant in the country, still in violation of federal standards for secondary treatment of sewage, and a source of bad odours in the district; 22 privately owned garbage transfer stations; over 200 sites that store or handle hazardous substances under the city's Right-to-Know law, (including 20 facilities reporting to the Federal Toxic Release Inventory); as well as the city's only radioactive storage facility. In addition, the district is also home to a great many smaller polluting industries that are not regulated. Also worth mentioning is the proposed super incinerator being planned for the Brooklyn Navy Yard, at the southern boundary of the district. The concentration of these problems make this district one of the most environmentally burdened in the city, and in the country.[2]

12.3   Community Activism and the Environmental Benefits Program

Responding to these problems, the community organised to protect its environment. Community groups, such as The Toxic Avengers of El Puente, Neighbors Against Garbage (NAG), or Concencerned Citizens of Greenpoint, were formed to deal with specific environmental concerns. Community activism also led to the unique Environmental Benefits Program (EBP) developed by and for the community. The City established EBP with $850,000 in response to a court settlement with the state of New York over violations of the Clean Water Act by Greenpoint-Williamsburg's sewage treatment plant. The program's objectives were to 'identify and assess environmental problems, initiate projects to reduce pollution and promote environmentally sound development' (NYC Department of Environmental Protection 1993; ICLEI 1993). A number of projects were undertaken through the program:

- A series of environmental health/education and environmental education/training programs;
- The Clean Industries Program, a technical assistance program to encourage businesses to invest in pollution-reducing technologies, and to attract green businesses to GW;

- The development of a GIS set of environmental and other data bases to enable the community to monitor its environment and keep track of the record of industries within the district;
- The adaptation of Integrated Environmental Zoning (IEZ) to enable the community to aggregate its many types and sources of pollution and provide a comprehensive baseline of its environmental conditions; and
- The establishment of the first Urban Environmental Watchperson's office - a community resource center with a full-time environmental advocate and computer access to the GIS information and the results of the IEZ adaptation.

12.4   The Baseline Aggregate Environmental Loads (BAEL) Project

The Baseline Aggregate Environmental Loads (BAEL) study was funded by the Environmental Benefits Program to build on the GIS set of data bases already developed for the area (Ahearn and Osleeb, 1993), and go beyond the piecemeal approach to environmental problems. It turned to the Integrated Environmental Zoning (IEZ) method of the Dutch Ministry of Housing, Physical Planning, and the Environment (VROM) to develop an aggregate environmental baseline profile of Greenpoint-Williamsburg (Anderson et al., 1997). As the lead urban planner on the research team[3] from Hunter College, which also included geographers and community health experts, during 1995-1996, I identified and developed the potential applications of the baseline profile.

12.5   Integrated Environmental Zoning

Integrated Environmental Zoning is the most innovative spatially oriented approach that the Dutch government proposed as part of its National Environmental Policy Plan of 1989 (Lloyd's Register, 1993; Sol et al., 1995; Miller and de Roo, 1996).

In effect, IEZ is a type of suitability analysis, with a methodology for aggregating many sources and types of environmental impacts in a geographic area into a set of scores. (McHarg, 1961; Hopkins, 1977; Westman, 1985) are then mapped and, as in suitability analysis, the resulting maps are then used for making land use and other policy decisions. IEZ focused on five environmental layers:

- air toxics – carcinogenic;
- air toxics – non-carcinogenic;
- hazardous materials;
- noise; and
- odour.

In IEZ, each environmental load category (index) is scored on a scale that ranges from acceptable (A) to unacceptable (E). When the loads in the five indices are combined, they yield a summary index of six 'integrated classes', I-VI, which identify the total environmental load at a location. Class 1 is deemed appropriate for residential areas. Classes II-VI allow some flexibility for residential uses, and Class VI is interpreted as having an environmental load greater than is allowed for residential use (VROM, 1990).

These summary classes, I-VI, trigger specific source pollution reduction strategies. If the pollution reduction program does not reduce pollution to an acceptable level, land use controls are then applied. Two different degrees of restrictions are applied depending on whether the facility or source in question is an existing or planned development. Source limitations and modification requirements are stricter for new development than for existing facilities. In an existing situation, more relevant to Greenpoint-Williamsburg's conditions, a class IV area classification triggers policy decisions that prohibit large increases in the number of inhabitants. A class VI designation based upon existing sources and facilities may trigger actual demolition of housing. In areas with the heaviest loads that contain housing, Dutch municipalities were required to prepare plans that specify abatement measures, and strategies to accommodate anticipated manufacturing without harming residents, including both pollution prevention and land use regulations (Miller and de Roo, 1996, pp. 376-7).

IEZ is an innovative and ambitious version of suitability analysis. Noteworthy is that the Dutch government developed this method to protect urban environmentally sensitive areas, viz., residential areas. This differs from the way that suitability analysis has been typically used in the United States, where the concept, 'environmentally sensitive areas' has been reserved for natural resource areas, such as wetlands. In the United States, suitability analysis itself has been used to determine the least environmentally disruptive location for a specific use, such as a highway or a new residential development. This use of suitability analysis *within* an urban context, focuses on the urban context's most valued resource – its residential neighbourhoods. This focus on the exposure of neighbourhoods to environmental impacts suggests a new name for this type of analysis – vulnerability analysis.

IEZ is also one of the most ambitious efforts to apply suitability analysis. 12 cities in the Netherlands participated in the IEZ experiments. The national resources spent in data monitoring, collection, and processing, and research to develop thresholds and limits, and credible weighing schemes represents a heroic effort to develop an aggregate environmental index.

12.6 BAEL Features[4]

In order to determine the feasibility of applying IEZ to GW, the BAEL team assessed the availability of comparable input data and models. The first decision involved the choice of the geographical unit of analysis. Given the rich set of data available from the US Census, already incorporated in the GIS developed for Greenpoint-Williamsburg, the choice was between census tracts and census block groups. The census block group was selected as the unit of analysis, of which there are 159 in Greenpoint-Williamsburg, since it enabled a more fine-grained analysis than census tracts. See Table 12.1 for a summary of BAEL features.

For the air toxics indices, both carcinogenic and non-carcinogenic, BAEL relies on two US Environment Protection Agency models, a dispersion model and the Environmental Indicators Model (Abt, 1992). As of summer 1997, due to resource and data accessibility constraints, the BAEL profile contained only the federal Toxic Release Inventory data – 20 reporting facilities. Federal regulations require all facilities that release over a certain amount of one or more of a list of several hundred toxic chemicals to report the amount to TRI. The releases reported by the TRI facilities are used as inputs for a dispersion model which calculates annual average chemical concentrations at the centroid of each block group. The dispersion of toxicant and carcinogenic releases are modelled using Industrial Source Complex Long Term (ISCLTR3) software available from EPA, a model which is used primarily for permit applications. The toxicant index for a block group is equal to the sum of its toxicity indicator elements. These elements are calculated according to a formula adapted from EPA's Environmental Indicators Model. In addition to the emissions from the 20 TRI facilities included in BAEL's air toxics index, plans are underway to add and model the releases of facilities requiring air pollution permits under the city's and state's regulations. This will add about 400 facilities to the index.

Table 12.1   BAEL indices and inputs

Environmental Layers/Indices	Inputs	
Air Toxics – carcinogens and non-carcinogens	Data:	TRI data.
	Modelling:	USEPA's Environmental Indicators Model, ISCLTR3 dispersion model.
Hazardous Materials	Data:	NYCDEP's Right-to-Know data.
	Modelling:	NYCDEP's MCP hazard ratings and a decay function.
Noise	Data:	NYCDEP's complaint reports, and traffic counts along truck routes.
	Modelling:	Traffic to noise conversion table.
Odour	Data:	NYCDEP's complaint reports.

Since comparable risk assessments of stored hazardous materials were not available for Greenpoint-Williamsburg, BAEL developed an index based on hazardous materials information available from New York City's Department of Environmental Protection. NYCDEP collects information on facilities that store hazardous substances under the City's Right-to-Know law. The data included 161 chemicals stored by 284 facilities in the district. This information includes the quantity of hazardous materials, as well as its Mass Casualty Potential (MCP) rating. The MCP is an expert but subjective assessment of a chemical's potential to cause public casualties and reflects a chemical's flammability, reactivity, and explosivity. The BAEL index on hazardous materials used an additive model that summed all the reportable quantities and weighed the quantities based on the MCP scores and the proximity of a block group to a reporting facility.

Unlike the Dutch government, the US federal government does not have federal regulations to control noise or odour. New York City, like many cities in the US, does have a noise code, but ambient noise levels are not monitored in NYC, and no noise measurements were available for Greenpoint-Williamsburg. The sources of the noise and odour indices in BAEL were complaints reported to NYCDEP by citizens. In addition, BAEL extrapolated the noise generated from trucks and traffic along major transportation corridors by using a traffic to noise conversion table developed by Harmelink (1970) and presented in the ITE Transportation and Traffic Engineering Handbook.

In seeking to adapt and then apply IEZ to this community in NYC, many data and methodological problems surfaced. The unit of analysis selected, the census block group, although the best available,[5] introduced validity errors in the attribution of scores. For example, for air toxics, the centroid of the block groups was selected as receptor sites for the dispersion model. Entire block groups are assigned a score based on that assumption. Since block groups vary in size and shape, scores are problematic for unusually large or irregularly shaped block groups.

The lack of monitored data on air pollution, noise and odour required us to rely on models, extrapolations, or unsubstantiated complaints. The lack of surveys on the perception and priorities of the Greenpoint-Williamsburg community, and the lack of standards on where to draw the limits and thresholds for some of the indices also dampen confidence on the weighing, scoring, and aggregation choices. As a result, a major problem we confronted was the lack of a credible basis for, and agreement on how to aggregate the various layers. BAEL faced many data problems, including:

- The incompatibility of data bases, e.g., some were aggregated by address and others by block and lot number;
- The lack of uniformity and co-ordination in reporting among different levels of government – for example, health districts and census tracts have no relationship to each other; and
- Problems with finding adequate data, e.g., lack of facility specific hazardous materials risk assessments.

Finally, the GIS and modelling covered only one out of 59 districts in the city, and, thus, the results are only comparative *within* the district.

Due to the problems cited above, the team agreed that aggregating the five indices into an integrated one, as in IEZ, lacked credibility. Some agreement was reached that if the data for hazardous materials used a more objective system than Mass Casualty Potential, then the air toxics and hazardous materials indices could be more credibly aggregated. In addition, if measurements for noise and odours were obtained, then these two indices could also be aggregated. Thus, if further developed along these lines, BAEL could contain two aggregate indices, one directly related to health and safety, and the other to nuisance.

12.7   Promising Applications

With all the caveats above, when combined, the GIS set of data bases and the BAEL modelling result in a set of maps that display the five indices or layers visually in such a way that their applications seem almost self-evident. These maps constitute a profile of the environmental vulnerability or sensitivity of residential areas within the district. Since this is a pilot project, and no comparable analysis exists for any other district in NYC, this environmental profile of Greenpoint-Williamsburg is an exploratory and suggestive comparison of the environmental conditions of different census block groups within the district. The set of maps can be used to inform various regulatory, planning and resource allocation processes. In the BAEL Final report I develop over two dozen recommendations for the use of the BAEL profile maps. In this section, for illustrative purposes, I will use the air toxics layer generated by BAEL to explain the potential application of this kind of spatial information in several planning contexts. Table 12.2 summarises some of the major applications.

Table 12.2   Potential applications of BAEL to types of decisions

Type of Decision	BAEL Application
Siting	In heavy exposure areas, BAEL maps can flag pollution prevention needs, or need for environmental assessment.
Zoning and Land-Use	Used as a factor to determine whether an area should be re-zoned, e.g. from industrial to mixed use or residential, or require a special permit; ideally, BAEL could be used to determine industrial zoning boundaries.
Environmental Regulations	More refined targeting of areas for greater scrutiny, and for prioritising enforcement or inspections; designating urban CEAs.
Resource Allocation	Used to prioritise budget allocation decisions.

Siting

Figure 12.2 presents the loading of one layer, non-carcinogenic air toxics and one type of source – TRI facilities by block group. As mentioned above, there are 20 TRI reporters in the district. Since there is no data on the concentration of these toxics in the air, BAEL used a toxicity scoring system that the USEPA developed, the Environmental Indicators Model, and then a dispersion model used by EPA. Within Greenpoint-Williamsburg, the map identifies areas with comparative greater or lesser loads of these TRI toxicants. As with IEZ, the lightest areas in the map, with scores of 1 or 2 show the least affected block groups, while the darkest, with scores of 4 or 5 indicate the block groups that have the highest exposures to the air toxicants.

Toxicity-weighted emissions

	0 - 15		18 - 106
	15 - 18		106 - 385

Figure 12.2 Index of exposure to airborne toxicants from TRI reporters – by block group, Greenpoint-Williamsburg

Suppose a new facility with significant air impacts, but below regulatory thresholds, is proposed. By locating the proposed facility on the map, we can visually determine whether the proposed location has a comparatively greater air toxics load than other locations. In a heavy exposure area, the BAEL Profile maps can be used in the environmental assessment process as a quick visual scanning tool to flag whether a development proposal should receive heightened attention or scrutiny, either through informal ways – by persuading the developer to incorporate pollution prevention technology in the project, or by initiating an EIS process. The methodology used in the air toxics index can also be used to estimate the increased overall load on a block group a new facility would generate. In NYC, in particular, the method can be used to introduce a more scientific approach to the City's Fair Share Policy (Weisberg 1993). The current policy uses a map of existing facilities to determine the fair distribution of regional/citywide facilities among the boroughs in the city. BAEL goes beyond the eyeball approach in determining environmental burdens and provides a way to aggregate the potential environmental impacts of multiple facilities at a finer grain.

Zoning and Land Use

A major, if not the major, purpose of zoning is to protect single family residential districts from nuisances, in particular, industrial nuisances. This is largely accomplished through zoning by restricting industries to certain areas. Zoning in this country has relied on two ways to reduce nuisances for residential populations: a) prescriptive measures, i.e., detailing the list of permitted uses in different districts; b) performance measures, i.e., instead of listing permitted uses, performance standards require that industries 'perform' to a certain level, by setting standards of performance, for example, noise threshold levels, which, if complied with, allow any type of industry to locate in a zone. New York City uses a combination of the two approaches, both aspects of which require updating and revision.[6] In general, neither the prescriptive nor the performance approach attempts to incorporate aggregate measures of the environmental impacts of multiple sources. Even performance standards are geared to limit the impacts of individual projects and do not take cumulative impacts into account. Moreover, even the most sophisticated prescriptive or performance approaches often fail to incorporate the best knowledge we have about pollution and hazards.

IEZ seeks to remedy these problems by using up to date methods and empirical data to reform zoning in the Netherlands. Even with our limited adaptation of IEZ, BAEL demonstrates the problems with current zoning. For example, take the issue of toxic air pollution. Air toxics are dispersed primarily from high stacks to surrounding areas. Zoning boundaries that seek to protect residential areas from toxic emissions should take into account the typical dispersion patterns in the surrounding areas. Air dispersion modelling of air toxics in BAEL results in a profile of the district where the census blocks with the heaviest toxic loads have no TRI facilities located directly in their geographic boundaries. If we compare the air toxics layer map to the zoning map, as in Figure 12.3, we find that the areas with the highest exposure to airborne toxicants are residential and mixed use areas, and, in general, that much of the area with the heaviest loads are residential areas. Zoning rests on a simple but wrong assumption that pollution and hazards can be contained within narrow geographic boundaries. Dispersion modelling of air toxics demonstrates graphically how traditional industrial zoning boundaries fail to protect public health and safety. The revision of zoning in this country, on the basis of modelling the dispersion of air pollutants and estimating the risk of hazards, may be slow given the very contrasting political attitudes toward environmental protection in the United States and the Netherlands. But if zoning is to protect people from industrial and other environmentally significant uses, then, in the future, it must turn to an approach similar to IEZ and BAEL.

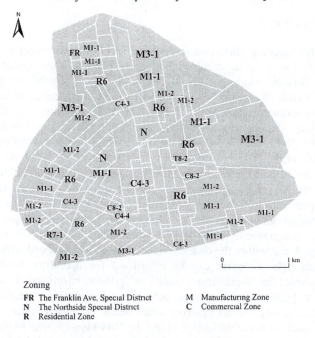

Zoning
FR The Franklin Ave. Special District   M   Manufacturing Zone
N   The Northside Special District   C   Commercial Zone
R   Residential Zone

Figure 12.3   Zoning map of Greenpoint-Williamsburg
Source: New York City Zoning Regulation Vol. II

The BAEL Profile maps can also be useful in reviewing zoning proposals. For example, NYC conducted a mixed use zoning study (NYCDCP, 1996) to determine whether the City should institute a mixed industrial/residential zoning designation that would permit industrial or residential development 'as of right' without substantive review. The BAEL profile maps suggest that in Greenpoint-Williamsburg, one of the few districts in the city where this proposed zoning designation could be applied, permitting residential or industrial uses as of right can either expose more residents to heavy environmental loads, or increase the environmental loads in already heavily loaded areas. Instead of permitting such projects as of right, special or conditional permits should be required. Such special permits should require measures to reduce existing environmental burdens and avoid additional ones. The City has also been studying changes in the zoning of community facilities, namely, it is considering permitting them in M-1, light industrial zones as of right (NYCDCP, 1993). The BAEL Profile indicates that some M-1 zones around the center of the district have heavy airborne toxicant levels. Air toxicants have greater effects on vulnerable populations such as children, the elderly, the sick, and, especially, those with immunity disorders. BAEL findings suggest that the location of facilities serving vulnerable populations in M-1 zones should require more rather than less scrutiny, and at the very least a special or conditional permit.

Resource Allocation

The vehicle for resource allocation at the local level of government is the budget process. NYC has a rather complex budgeting process with opportunities for Community Boards, Borough Presidents, and city agencies to participate. Community Boards, for example, prepare and submit a statement of their budget priorities and recommendations to the Mayor and city agencies. Community Boards also have the opportunity to comment on the Mayor's preliminary budget statement. The BAEL profile provides measures within the district of comparative environmental loads. These comparative ratings could he used by DEP, the Community Board and the Borough President, who play important roles in the budget process, to set budget priorities for the allocation of city funding for projects and programs to reduce the environmental burdens in the district.

The profile can be employed by DEP and the Community Board to develop a consistent, comprehensive set of environmental priorities for Greenpoint-Williamsburg. Areas within the district which are heavily burdened with air toxics, hazardous materials, or noise can be prioritised for operating funds that, for example, provide for more inspections, or greater enforcement of existing regulations, or technical assistance to industries to prevent or reduce pollution. Such areas can be targeted for environmental health education programs, and for capital items such as noise mitigation along the Brooklyn-Queens Expressway, or air pollution mitigation along major roads. When combined with health statistics, the profile can also be used to pinpoint residential areas with vulnerable populations, such as areas with a high incidence of respiratory illnesses, which have heavy air pollution exposure. In such cases, BAEL can be used to make the case for the location of new facilities, such as a new health clinic.

Environmental Regulations

Beyond the siting of facilities under the EIS process, there are many permits and regulations that require the reporting of activities that generate pollution or hazards. One of the complicating factors in trying to identify the potential applications of BAEL to environmental regulations is that these permits and regulations often occur at three levels of government, federal, state, and local. In addition to the federal TRI program, the federal Clean Air Act gives states and localities the authority to regulate the emissions of air pollutants. The New York State Department of Environmental Conservation and NYCDEP are responsible for issuing permits for all equipment emitting pollutants into the air, such as boilers, dry cleaning, gasoline dispensing sites, or auto body refinishing operations. Each company is required to report the quantity of toxic chemicals it discharges into the air. All emission-producing equipment requires a certificate from DEP to operate or a registration, which must be renewed every three years. Before a certificate to operate is issued, the equipment must pass an inspection.

Due to lack of sufficient personnel and other problems, audits of DEP's Air Permit Inspection programs found that the average response time for an air pollution complaint was 21 calendar days, and that inspections often took place a year or more after the certificate to operate expired. Given these operating delays, the BAEL profile maps can be used as a screen in permit procedures to alert inspectors to the sensitivity of the area in which a facility proposes to operate. In general, in an area heavily burdened with air toxics or hazardous materials, inspectors could be directed to prioritise inspections or to refer applicants to a pollution reduction technical assistance program.

BAEL is a systematic attempt to apply acceptable methods of aggregating the best knowledge and data available on environmental issues of concern to local communities. Some of the findings of BAEL, even at this pilot stage, throw into question existing standards used in the regulatory process. For example, exempting highways and traffic form the calculation of ambient noise quality, as NYC's current Noise Code[7] does, is likely to make the enforcement of such a code in residential areas of Greenpoint-Williamsburg, which are impacted by the Brooklyn-Queens Expressway and truck traffic, inconsequential. As BAEL estimates, if traffic alone brings the ambient noise quality to 65dB(A) and above, which is the maximum dB(A) permitted for residential areas in the district, the current standard is regularly exceeded in some parts of the district (see Figure 12.4). This finding suggests that BAEL can become a significant tool in testing the validity and usefulness of existing standards.

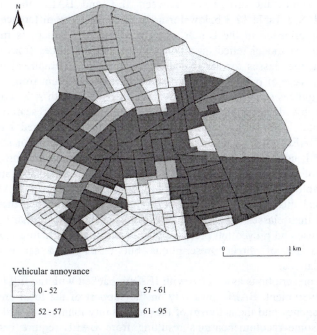

Vehicular annoyance

- 0 - 52
- 52 - 57
- 57 - 61
- 61 - 95

Figure 12.4   Index of vehicular noise annoyance – block group and truck and elevated train routes, Greenpoint-Williamsburg

The BAEL profile of Greenpoint-Williamsburg, which graphically displays a community overburdened with environmental problems, suggests a final recommendation. This recommendation is inspired by a major contribution of the Dutch model, i.e., the recognition that neighbourhoods are 'environmentally sensitive areas', and should be accorded similar protections as natural resource areas. New York state, like most states and the federal government, includes in its Environmental Quality Review legislation, a clause that allows the local designation of Critical Environmental Areas (CEAs). This designation has been primarily applied to natural resource areas, but this type of legislation can be interpreted to include the designation of environmentally burdened urban areas, such as Greenpoint-Williamsburg.[8] The benefit of such a designation for districts such as Greenpoint-Williamsburg is that the siting of uses with significant environmental impacts could be reduced, since the CEA designation triggers the environmental assessment of all major projects. In combination with an aggressive pollution reduction program for existing uses, this designation could become a comprehensive way of dealing with the multiple problems in communities suffering from environmental injustices.

12.8   Differences between IEZ and BAEL

There are significant differences between IEZ and BAEL that need to be emphasised. See Table 12.3 below for a summary of these differences. First, the degree of confidence in the Dutch provisional model's indices is much higher, since the Dutch model relied on ample support and resources from the national government, the classes for the five indices are based on empirical data, and on extensive surveys of public perception and priorities. Different from IEZ, none of the BAEL indices stand for comparative health risks. Health risks could only be established by documenting potential exposure and by research that links environmental loads with health outcomes. Due to the strained nature of our resources, BAEL relies on reports of complaints and on models for several of the indices, and conducted no surveys to establish community priorities. The developers of IEZ were able to establish clearly what was acceptable and what was not acceptable for each layer, and obtain government support. BAEL has not accomplished this for most layers. The high level of confidence that the Dutch secured for their classes, enabled the direct linkage of each summary class with a set of land use and prevention prescriptions. Due to its pilot nature, BAEL cannot result in a set of direct prescriptions but rather in a set of suggestive recommendations.

The prescriptions associated with IEZ are backed with the full power of the national government. BAEL must rely on the support of the local environmental protection agency and the activism of the Community Planning Board and district residents. Some recommendations resulting from BAEL require resources that must be won in the competitive arena of the budget process. Others require more or less extensive regulatory changes in city procedures. The value of BAEL for Greenpoint-Williamsburg is its ability to bring together in a systematic and scientific way the various levels of pollutants, hazards, and nuisances that plague the district. BAEL is a tool that both DEP and the Community Board can use to make this environmentally sensitive district more sustainable in the years ahead.

Table 12.3   Comparison of VROM's IEZ to Greenpoint-Williamsburg's BAEL

Features	VROM's IEZ	GW's BAEL
Context	National government policy	Relies on community activism and support by local government
Inputs	Data on air pollution, hazardous materials, noise and odour based on solid measurements as well as surveys of public perception and priorities	Some data on air pollution and hazardous materials, but self-reported; no measured data on noise or odour; no surveys of public perception or priorities
Outputs	Health-related toxicity ratings and risk assessments by geographic unit	Exposure or potential environmental impacts by geographic unit
Policy Use	Prescriptions on source reduction and land uses	Agenda for community action

12.9   The Broader Significance of IEZ and BAEL in the US Context

What is IEZ's and BAEL's significance for planning? Does IEZ deliver its promise of providing a cumulative environmental baseline? What about BAEL's significance for its two clients, NYCDEP and the community of Greenpoint-Williamsburg? Is BAEL a model for community organisations interested in incorporating information technology into their operations?

BAEL has two clients, NYCDEP and the community of Greenpoint-Williamsburg, as represented by a Citizens Advisory Committee. What is BAEL's significance to these two clients?

For NYCDEP, and local environmental departments in general, the use of GIS-based environmental indices has great potential for incorporating environmental concerns early on in all the relevant physical planning and development processes in a city. The further development and upkeep of these environmental layers is in the self-interest of local environmental departments. The GIS-generated maps, once developed for the entire city and required as inputs in relevant local government processes, extend the influence and presence of local DEPs. The maps, in effect, serve as quick scan mechanisms to ensure that environmental concerns are represented in all relevant discussions, even when DEP staff is not in attendance. Despite the many data problems yet to be overcome, BAEL-like, GIS-based environmental indices, with greater or less sophistication, will become as central to local DEPs, as existing land use plans are to city planning departments.

How does BAEL measure up to its goal of providing an interactive, citizen-friendly system for communities to use in protecting and improving their environmental quality? The answer to this question is more nuanced. The project's goal of developing a GIS-based environmental profile of the community with specific recommendations for its use has been accomplished. But what is BAEL's broader significance for communities seeking information technology to monitor and assess environmental quality? Is BAEL a model to be emulated? Several barriers stand in the way of an affirmative answer. These are: data requirements, GIS technology, and institutional context.

The Environmental Benefits Program was developed by a set of very committed and smart planners and community activists who included as part of the program the establishment of a non-profit community institution, the Environmental Watchperson's Office to house the environmental advocate and the GIS. The Environmental Watchperson's Office has been in existence since 1995, but its future is in question, since EBP funds have been spent. Communities interested in the BAEL approach need to first secure a stable institution to maintain and provide community access to the GIS. A second barrier is the technology itself. GIS software requires special training. In the case of BAEL, maintaining and accessing the GIS required a part-time staff member trained in GIS. A most formidable barrier for communities interested in doing a BAEL-like study is gaining access to and processing the data needed as inputs for the system, as well as the analysis and modelling. In particular, community organisations, on their own, and even with local government support, are not likely to have comparable access to data. And most community organisations do not have the resources to hire staff or consultants to develop a system comparable to BAEL. Even for communities with the resources, it does not make sense to develop environmental indices on their own. In order for environmental indices to be influential in planning decisions, they should be developed in partnership with local government. BAEL has the most promise for community use in cities with strong and recognised neighbourhood associations and neighbourhood planning capacity. It is in such offices, already staffed with planners trained in GIS, and with open access to data sources, that environmental indices can be developed and applied for the benefit of neighbourhoods.

Is IEZ the way for zoning in the twenty-first century? IEZ is not perfect. Even buttressed with surveys of residents' perceptions and priorities, the aggregation of such disparate layers as carcinogenic air toxics and odours into a single index is problematic. The prescriptive policy use of the single index for determining public decisions seem inflexible, even to the Dutch cities that have been experimenting with IEZ (Miller and de Roo, 1997). The resulting environmental profiles of the Dutch cities participating in the IEZ experiment has been alarming. Almost entire cities appear blacked out in IEZ maps indicating that if source reduction policies are insufficient to reduce environmental loads in these cities, then no new development should be allowed and even that existing neighbourhoods or industries should be relocated. This has led the Dutch government to reconsider IEZ as their preferred approach, and various new experiments are being pursued, including pollution trading systems (Sol et al., 1998), and devolution of power to localities to allow greater policy flexibility (van Staalduine and Simons, 1998; Miller and de Roo, 1998; Blanco, 1998). Of particular relevance to BAEL is the environmental matrix approach in Amsterdam (Groot and Vermeulen, 1997) which uses GIS produced maps of environmental layers and other planning factors to inform the planning process. In this approach, the environmental layers are used as input factors in planning decisions, instead of within the prescriptive framework of IEZ in which the mapping triggers specific policies. This more flexible, descriptive, and disaggregated use of some of the five indices resembles the approach I developed for the BAEL study. To my mind, this is also BAEL's major contribution to urban planning. I believe that due to data constraints, unresolved methodological problems, and lack of political will, we are still far from the Holy Grail of a cumulative environmental index. BAEL, however, does demonstrate how GIS-based modelling can generate environmental indices that can bring land use planning and environmental policy into closer co-ordination. These types of urban environmental overlays can he developed and used in siting, land use planning and zoning, environmental permitting, and the budget process to approximate a more integrated urban environmental management. This outcome is vitally important to all urban residents in an increasingly urban world. It is of particular importance to poor or working class communities such as Greenpoint-Williamsburg which bear disproportionate environmental burdens.

Notes

1   Hilda Blanco is Professor at the Department of Urban Design and Planning, University of Washington, Seattle, USA.
2   This community profile draws from *Bridges: A Comprehensive Waterfront Plan for Greenpoint-Williamsburg,* prepared by the Hunter Planning Studio, Spring 1995, which I led.
3   The included Profs. Jeff Osleeb and Sean Ahern, and Anthony Baimonte from the Geography Department; Profs. Steve Zoloth, and Dan Kass, and Darius Sivin from Center for Occupational and Environmental Health, and myself and Prof. William Milczarski from the Department of Urban Affairs and Planning, Hunter College.
4   This section and the following section on applications draws heavily from BAEL's Final Report to NYCDEP.
5   See BAEL Final Report, 1997.
6   See BAEL Final Report, 1997. Research on the NYC industrial zoning regulations conducted for BAEL revealed that both prescriptive and performance requirements were obsolete, and in need of revision.
7   NYCDEP March 1992, 17.
8   One criterion for CEA designation under NYSEQRA is that a property presents 'a benefit or threat to the public health or public safety'. The examples of threats to public health or safety in the SEQRA Handbook (NYSDEC 1982, B-2) are of an abandoned land-fill or of a flood hazard area. Although to date, CEAs have not been designated to protect urban residents from a concentration of environmental impacts, SEQRA's criteria for designation may be open to such a use.

References

Abt Associates (1992) *Toxics Release Inventory: Environmental indicators methodology*, Abt Associates Inc, Bethesda.

Ahearn, S., and J. Osleeb (1993) Greenpoint/Williamsburg environmental benefits program: Development of a pilot geographic information system, *Proceedings of GIS/LIS*, Nov 1993, pp. 1-12.

Anderson, N., E. Hanhardt and I. Pasher (1997) From measurement to measures, in D. Miller and G. de Roo (eds) *Urban Environmental Planning*, Avebury, Aldershot, UK.

Blanco, H. (1998) A Unites States perspective on the Dutch government's approach seeking greater cohesion in environmental and spatial policy, in D. Miller and G. de Roo (eds) *Integrating City Planning and Environmental Improvement*, Ashgate, Aldershot, UK.

Dutch Ministry of Housing, Physical Planning and the Environment (VROM) (1990) *Ministerial Manual for a Provisional System of Integral Environmental Zoning*, Dutch Ministry of Housing, Physical Planning, and the Environment, The Hague.

Groot, M.M., and J.W. Vermeulen (1997) The environmental matrix enhances planning processes, Unpublished paper presented at the Second International Symposium on Urban Planning and Environment held in Groningen, The Netherlands, March 1997.

Hopkins, L.D. (1977) Methods for generating land suitability maps: a comparative evaluation, *JAIP*, Vol. 43, pp. 386-400.

Hunter College Planning Studio (1995) *Bridges: Greenpoint-Williamsburg Waterfront Plan*, Department of Urban Affairs and Planning, Hunter College, CUNY.

International Council for Local Environmental Initiatives (1993) Community-Based Environmental Management, New York City, USA, Case Study #14.

Lloyd's Register (1993) *A Study on Environmental Zoning Systems in 12 Industrialised Countries*, The Dutch Ministry of Housing, Physical Planning and the Environment, The Hague.

McHarg, I. (1969) *Design with Nature*, Doubleday/Natural History Press, New York.

Miller, D., and G. de Roo (1996) Integrated Environmental Zoning, *JAPA*, Vol. 62, No. 3, pp. 373-380.

Miller, D., and G. de Roo (1997) Transitions in Dutch environmental planning: new solutions for integrating spatial and environmental policies, *Environment and Planning B: Planning and Design*, Vol. 24, pp. 427-436.

NYC Department of City Planning (1993) Community Facilities Zoning Study.

NYC Department of City Planning (1996) Zoning to Facilitate Housing Production

NYC Department of Environmental Protection (1992) *Noise Code*.

NYC Department of Environmental Protection (1993) Environmental Benefits Program, *Newsletter*, Office of Community Environmental Development.

NYC Department of Environmental Protection (1996) Environmental Assessment Statement: Mixed Use and Related Zoning Text Amendments.

NYC Department of Environmental Protection (1997) *Baseline Aggregate Environmental Loads (BAEL) Project for Greenpoint-Williamsburg, Final Report*.

NYC Office of the Comptroller (1994) Audit Report on the Department of Environmental Protection's Air Pollution Inspection Program.

NYC Office of the Comptroller (1996) Follow-Up Audit Report on the Department of Environmental Protection's Air Pollution Inspection Program.

NYS Department of Environmental Conservation (1982) *The SEQR Handbook*.

Sol, V. M., P.E.M. Lammers, H. Aiking, J. De Boer and J.F. Feenstra (1995) Integrated Environmental Index for Application in *Land Use Zoning, Environmental*

Management, Vol. 19, No. 3, pp. 457-467.

Sol, V.M., J. de Boer, F.H. Oosterhuis, J.F. Feenstra and H. Verbruggen (1998) The city bubble: a framework for the integration of environment, economy, and spatial planning, in D. Miller and G. de Roo (eds) *Integrating City Planning and Environmental Improvement*.

Staalduine, J.A. van, and M.T.T Simons (1998) Environment and Space: towards more cohesion in environmental and spatial policy, in D. Miller and G. de Roo (eds) *Integrating City Planning and Environmental Improvement*.

Weisberg, B. (1993) One City's approach to NIMBY: How New York City developed a Fair Share Siting Process, *JAPA*, Vol. 59, No. 1, pp. 93-97.

Westman, W.E. (1985) *Ecology, Impact Assessment and Environmental Planning*, John Wiley and Sons, New York City.

Chapter 13

Urban Development and the Role of Strategic Environmental Policy Planning: Experiences with the First Generation of Plans in The Netherlands

F.H.J.M. Coenen[1]

13.1 Introduction

Many municipalities all over the world have adopted environmental declarations, strategies and action plans. A general term for these plans that address environmental problems is 'green plans' or 'environmental (policy) plans'. The relation between local physical, land-use or spatial planning and these forms of environmental or green planning varies from system to system. In some countries the planning functions are combined while in other countries they are quit separate (Sustainable cities project, 1994).

In this chapter we discuss the role that separate local environmental plans can have in influencing environment-related urban decision making and planning. We will focus on environmentally relevant urban development decision making and policy planning like the recruitment and location of businesses, sustainable housing, traffic and physical planning. The Netherlands present an interesting case for two reasons. Firstly because the initial Dutch National Environmental Policy Plan (NEPP), published on May 25, 1989, is internationally seen as a successful forerunner for similar documents published in other countries. Dutch municipalities were also early involved in green planning exercises. Secondly because in the Dutch planning framework there are two separate tracks of environmental and physical planning, and the environmental planning system has an interesting philosophy concerning how environmental plans should influence decision making. The chapter does not discuss the relation in general between environmental policy and urban development, but very specifically the relation between the environmental plans and urban development decision making and planning. The central theme of the chapter is the use of these environmental policy plans in decision making. The three main research questions in this chapter are: How do environmental policy plans influence urban development decision making in theory and in practice?; What are obstructing and facilitating circumstances for the use of these plans?; To what extent does this influence depend on the form of planning?

The conclusions of the role of environmental policy plans in urban planning are based on an evaluation of the experiences with the first generation of municipal environmental policy plans in The Netherlands. Central in this evaluation was the so-called ex-post linked decision making. Ex-post linked decision making is environmentally relevant decision making which takes place after the plan has been approved, and which is influenced by the plan.

13.2   The Dutch Planning Framework

Table 13.1 gives a simplified presentation of the Dutch environmental and spatial planning framework. It shows two different tracks of policy planning at three levels of government which are historically related to the environment, but based on different planning law, namely a law on physical planning (*Wet op de Ruimtelijke Ordening*) and a law on environmental protection (*Wet Milieubeheer*) which since 1993 contains a Planning Chapter. At all three levels some form of environmental and spatial plans have to be made, although the environmental plan on the municipal level is facultative.

Table 13.1   The Dutch environmental and physical planning system

Level of government	Environmental planning	Spatial planning
National	National Environmental Plan	National Spatial Plan
Provincial	Environmental Policy Plan	*Streekplan* (provincial land use plan)
Municipal Level	Environmental Policy Plan	Structural plan and *Bestemmingsplan* (local housing and landscape plan)

The table is overly simplified because it leaves out the specific operational programs that have to be made at the all three levels. On the municipal level the environmental program is required. Additionally, other sector plans, like water and nature conservation plans on the national and provincial level and municipal sewage plans address contain parts of the environmental planning track.

In the environmental track there is no formal hierarchical co-ordination mechanism. There is also no formal horizontal mechanism in Dutch law to co-ordinate environmental and physical planning on the local level, although there are co-ordination mechanisms on the operational or permit level. The role that physical plans can play in environmental protection is well defined in law and jurisprudence. The prime objective of the physical plan is 'good physical planning', which restricts the possibilities for conducting environmental policy through physical planning.

Planning Philosophy

In this chapter we see planning as a special form of policy. Planning distinguishes itself from normal policy in that it is more focused on the connection between decisions, and is more oriented toward the future. The *raison d'être* of planning lies in the advantage of planning, in the opinion of policy-makers, as compared to ad-hoc decision making. The advantages of environmental planning are contained in a number of objectives that policy-makers at the government level aim to achieve through environmental policy planning. The government's central objective with regard to environmental policy planning reads that planning has to result in a higher quality of decision making. The objectives concern the features which decision making should have after planning. Motives analysis revealed the following objectives of environmental policy planning, as features of planned decision making (Coenen, 1996): taking future effects more into account in decision making; providing more insight into the future effects of decision making; co-ordinating a decision better with other decisions of the planning subject; providing other actors with more opportunity to influence a decision; and providing information on the actions of the planning subject and provide better arguments for decisions. For the relation between environmental policy plans and urban development the achievement of the objective mutual coherence between environmental and development decisions will be the main focus in this chapter.

Within the new policy planning framework, a lesson has been drawn from experiences in physical planning. A comparative study on physical planning in The Netherlands and Great Britain (Thomas et al., 1983) showed that municipal physical plans had little effect on the shape of the built environment. The original plan was constantly deviated from. The researchers advocated municipal physical plans which have less pretensions of determining future spatial planning.

In his memorandum 'More than the Sum of its Parts', the Minister of Housing, Physical Planning and Environment defines planning as: 'the development and maintenance of Statements of Future Intent in the form of plans to be able to take future decisions rationally and in their mutual context, and to convince third parties (other authorities, enterprises, citizens) to take this into account in their actions and decisions'. The ideas from this memorandum were converted into legislation in the chapter 'Plans' of the General Environmental Conservation Act. The concept of a plan as 'a statement of future intent for future decisions' and planning as 'the development and maintenance of statements of future intent to be able to take future decisions rationally and in their mutual context' places the concept of planning indicated in the memorandum 'More than the Sum of its Parts' within the so-called decision-centred view of planning (Faludi and Mastop, 1982; Faludi, 1987). The difference between planning and 'ordinary' decision making in the decision-centred view of planning is based on work by Friend and Jessop (1968), with planning recognising the uncertainties regarding choices or decisions which can be taken in future. Decision making becomes planning if the problems of choice that arise are connected with other choices that are related to them. The importance of planning is that it provides 'a guideline for future decisions' (Friend and Jessop, 1968, p. 111). Here, a plan is 'a statement of future intent'.

Directly opposed to the idea that plans are statements of future intent is the thought that a plan is something that has to be executed. In their most extreme form we speak of blue-print plans. Blue-print planning aims at determining and executing a desired final condition. In blue-print planning, the uncertainties with which we are confronted in planning are insufficiently taken into account, and this is precisely the explicit principle on which the decision-oriented approach to planning is based. From the decision-centred view, Faludi has argued that different criteria should be used for strategic and blue-print plans (project plans). The conformance view means the measuring of outcomes to intentions. The alternative to the conformance view is the 'performance view' (Barrett and Fudge, 1981). A strategic plan gives guidance. If a strategic plan is abandoned, this does not mean that it did not work. We have to look at the usefulness of the plan to the decision-makers.

A completely different concept of the impact of plans is the view held by Wildavsky (1973) and his followers. To Wildavsky planning is only a matter of belief. Wildavsky's conclusion reads that what we call planning cannot really be distinguished from ordinary ad hoc decision making by the authorities, which is not aimed to be planning. The hypothesis put forward by Wildavsky is that the effects of an unplanned policy and unplanned decisions do not differ from the effects of a planned policy and planned decisions. Planning is just a matter of belief.

Based on the general planning philosophy plans are given in the planning system two major functions. The *internal function* concerns the guiding of future decisions taken by the planning subject itself. The *external function* is the consequence of publicising the plan. By publicising policy intentions, actors in the environment such as citizens, firms, social organisations and other authorities gain an insight into the type of behaviour which is to be expected from the municipalities, so that they can adjust their own decision making to this.

The environmental planning system asks on the municipal level for the presence of a strategic environmental policy plan (facultative) with an accompanying implementation programme, or a plan that consists of both a strategic and an implementation part, aimed at the planning functions from the planning system. In its basic concept, the model of the planning system, i.e. the ideas about the organisation of the planning system and the course of the policy processes which have to take place within the planning system, dates back to 1984; it was given its final form after the publication of the bill for the Planning chapter (1989) and after this bill was discussed in parliament (June 1991). So although not formally developed under the planning law, all case-municipalities we will refer to in this chapter were influenced by the planning framework.

13.3   The Influence of Environmental Policy Plans in Theory

The first research question addressed is: how do environmental policy plans influence decision making in theory? The environmental planning system is based on the decision-oriented planning approach: a plan is a guideline for future decision making.

If a strategic plan is abandoned, this does not mean that the plan did not work. We have to look at the usefulness of the plan to the decision-makers. This approach results in assumptions on the way in which effective plans have an impact. Also the alternative approaches contain assumptions about the relation between the plan and ex-post decision making.

These different approaches have consequences for the function attributed to a plan and the intended result pursued through planning. The aim of blueprint planning is plan realisation, while the function of the plan is to indicate the final picture that is to be realised. The aim of decision-oriented planning is to improve the quality of decision making; here the function of the plan is to direct decision making. In the 'planning-as-a-religion' approach, planning does not pursue any effects, but the plan has latent or political functions. On the basis of each of the three alternative approaches an ideal-typical theoretical model may be constructed of the relation between the plan and the ex-post linked decision. The core of each theoretical model is an effect pattern which describes a sequence of events and/or actions between plan and ex-post linked decision. By an effect pattern we mean here a sequence of events in the relation between plan and decision making on the basis of the plan. The main elements for how a planning subject handles a plan consist of:

1   taking knowledge of the plan;
2   considering the plan relevant for the concerned ex-post linked decision making;
3   the consultation of the plan; and
4   using the plan.

If the planning subject is familiar with the plan, consults the plan and decides in accordance with the plan, it is called a *conformity effect pattern*. If the planning subject is familiar with the plan, considers the plan relevant, consults the plan and uses the plan explicitly in justifying his decision, it is called a *performance pattern*. If there is a lack of knowledge of the plan *or* not considering the plan relevant although it was relevant, *or* not consulting the plan although the decision gave reason to do so, *or* if the plan is deviated from without any argumentation it is called a *no-effect pattern*.

If we discuss the use of plans we can distinguish between four decision situations. A decision could conform without a reference to the plan, which could be accidentally conforming; be a deviation from the plan without explicit argumentation; conform with reference to the plan; or be an argued deviation from the plan.

13.4   Structure of the Study and Data Collection

The core of the research design consisted of a multiple-case study, consisting of seven municipalities each with more than 30,000 inhabitants and each with their own plan. Within these cases the environmentally relevant decisions with respect to the plan constitute the points of measurement. In addition, research was performed into the implementation of the environmental plans of all 110 municipalities with more than 30,000 inhabitants.

The decision research in the case-studies was based on two different data-collection methods, indicated as quantitative and qualitative decision making studies. The quantitative decision making study consisted of a contents analysis of decision making documents of the mayor and aldermen. To this end decisions from three research periods of 3 months each were studied for each of the case municipalities. The first research period ended about one year before the plan was approved. The other two research periods began eight and sixteen months, respectively, after approval of the plan. All in all, 511 decisions were judged, distributed over the seven municipalities. The qualitative decision making study consisted of reconstruction interviews with officials in the seven case municipalities. The interviews were supplemented with the contents analysis of the decision making documents.

For the relation between urban development and environmental policy plans, the relevant areas of research were: the relation with physical plans; the recruitment of new firms and location of new firms and services; traffic planning to reduce car-use and traffic nuisance and sustainable housing. We also looked into the relation between municipal policy and subjects that are of less interest for urban development like good-house keeping, the purchase of municipal goods like cars and the fighting of slipperiness and weeds.

13.5   Research Findings

Coherence between Environmental Policy Planning and Urban Planning Decision Making

One of the objectives of environmental planning is taking future decisions more in coherence. Target achievement would mean that urban development decisions, like for instance the recruitment and location of new firms, would be taken more in coherence with environmental decision making. The cause of this raise of mutual coherence could be the environmental plan, but also other factors like national environmental policy.

In the quantitative study a higher level of coherence was operationalised using the indicators 'the number of external integration decisions', 'the involvement of other services in decision making', and 'the number of decisions taken at the initiative of other services' as compared to the initial situation. The quantitative study showed only a very limited raise in the mutual coherence.[2]

In the qualitative part of the study coherence was operationalised as the question whether the ex-post linked decision making is more in coherence with environmental policy. 'More coherence' can be operationalised as being related mainly to the contents of decision making (content coherence), or being related mainly to its process (process-type coherence). For some types of urban decision making in the past, no explicit relation was made between this form of decision making and environmental planning, or there was a relation in the content but the environmental department was not involved in the decision making.

To answer the question of whether there was a growing mutual coherence, civil servants in other departments (like physical planners involved in the drafting of physical plans, civil servants involved in business recruitment, traffic planners and civil servants from the building department) were interviewed. In addition documents were studied, for instance newer plans were compared with older plans and specific cases, for instance the location of business, were compared between different time periods.

Mutual coherence can be observed between plans, separate strategic decisions and operational decisions. Coherence between plans especially plays a role in physical and traffic planning and sustainable housing. In the main municipal physical plan, the *bestemmingsplan* (local housing and landscape plan), there is an obligation to involve acoustic and soil research in the planning. More and more municipalities make separate 'environmental sections' in their physical plans. In all case-municipalities where there was a greater involvement of the environmental department in physical planning, the department commented on the environmental section and delivered the acoustic and soil research.

In traffic planning there is a national tendency for more coherence between traffic and environmental policy. We have to distinguish between traffic planning aiming at reducing car-mobility and aiming at noise nuisance. Some of the case municipalities were pioneers in car reduction, while others had a sole facilitating traffic policy concerned with leading the traffic streams. In the field of noise nuisance the larger municipalities were stimulated to make traffic audit plans. The process of making these plans stimulated the process coherence between the environmental and traffic department.

Concerning location policies and sustainable construction we may say that there is more coherence, where formerly there really was hardly any coherence. In the recruitment and location of firms the overall economic growth target of municipalities is of major importance for the way these municipalities handle environment. Municipalities with a large growth target, where several thousands of new jobs have to be created and thousands of houses have to be built with the necessary infrastructure, showed a more reserved environmental policy. There are even examples of municipalities trying to set the national environmental policy aside, like the Dutch ABC-location policy, on behalf of economic growth. On the other hand, there are municipalities with a low growth rate, and fewer employment problems who advertise themselves as 'green cities'. They are able to make choices in attracting new businesses and refuse environmentally unfriendly businesses. These are often municipalities which are very popular as central locations for offices.

In content coherence between the individual location and recruitment of firms and environmental policy there was already much coherence. What is changing is process coherence because environmental departments are more, earlier and more structurally involved in location and recruitment of business. If we look into building programs we see that the environmental conditions do not go beyond national policy. Most examples of sustainable building are specific model projects. Changes in the environmental conditions for building often took take place when there is a large housing development.

The qualitative analysis shows that in most of ex-post linked decision making there is more of a relation with environmental policy, because more systematic attention is paid to environmental issue (like in traffic policy and building) and because decision making takes environmental aspects into account at an earlier stage and in a more systematic way. There is more external integration, implementation of environmental policy through the other policy areas.

More coherence does not mean that the environmental plans caused this coherence. To give an insight in this question we have to look into the influence of the plans on decision making. Were the plans really used by the other sectors? And according to which patterns?

Conclusions on the Occurrence of the Patterns

In section four we stated four decision situations in relation with the plan. The quantitative research showed that in 39.4 per cent of ex-post linked decision making it could be established that the contents of the decisions were according to the plan (conformity). In 2.2 per cent of decision making it was found that decision making explicitly did not correspond to the plan. In a considerable percentage of ex-post linked decision making, i.e. 58.4 per cent, correspondence to the plan could not be established. This means that on the basis of the wording of the plan and the decision it could not be determined whether or not the decision corresponded to the plan. Only a small percentage of decisions, 10.9 per cent, referred explicitly to the environmental plan. In a small percentage of decision making (2.8 per cent), where there was reference to the plan, conformity could not be established.

The quantitative decision making study shows that we rarely see a performance pattern. A performance pattern is present when the decision situation equals 'conformity with reference to the plan' or 'argued deviations from the plan'. This means explicit written references to the environmental plan in the resolutions of mayor and aldermen are rare. In other words, the environmental plans play a limited part in the explicit justification of decision making. Because explicit justification is lacking, it is difficult to determine whether the similarity between plan and operational decision was not based on coincidence. This is only possible when there is explicit mention of the plan during decision making.

In a reference to the environmental plan in the sense of an explicit reference to a concrete plan statement, the performance pattern largely corresponds to the conformity pattern. The difference lies in the intention with which the plan is used: as a realisation of a previously articulated plan statement to be performed, or to support an ex-post linked decision. After all, the performance of a plan statement may be the main argument in favour of a certain ex-post linked decision.

The picture that explicit justification is rarely found is confirmed by the qualitative analysis. The no-effect pattern was operationalised by us on the basis of conditions such as the lack of knowledge of the plan *or* not considering the plan relevant although it was relevant, *or* not consulting the plan although the decision gave reason to do so, *or* if the plan is deviated from without any argumentation. The analysis shows that many respondents do not know the plan in the sense that they are unable to mention any concrete plan statements or do not consider the plan relevant to their choice situation for various reasons. The plan is rarely consulted. It appears that ex-post linked decision making largely would have taken place in the same way if no environmental plan had been present.

The conformity pattern appears to depend on the type of plan statements, and occurs particularly in the case of certain plan statements in which certain actions and measures were recorded in the plan. On the basis of the plan statement, the action is then initiated. The 'performance' effect pattern is rare. The impact study does show that much of the decision making takes place according to the plan, but explicit references to the environmental plan are rare. Performance was found mainly in ex-post linked strategic decision making.

13.6   Obstructive and Stimulating Factors in the Use of the Plans

The second research question posed earlier is: Which factors obstruct or stimulate the contribution of municipal environmental planning to urban development decision making? The stimulating and obstructive factors were arranged on the basis of their influence on key elements of the effect patterns distinguished in the theoretical models, i.e. knowledge of the plan, the plan's relevance to ex-post linked decision making, consultation of the plan and use of the plan.

Basically, all respondents were aware of the existence of an environmental plan within their municipalities. However, clear differences were found as to their knowledge of the contents of the plan. Respondents were asked to mention plan statements in their own policy fields. On the basis of the qualitative analysis the differences in knowledge of the plan between municipal respondents appear to be due to three factors: the organisational form of the municipality, the planning process followed, and the policy field.

From the qualitative analysis various factors emerged which made the environmental plan not *relevant* to ex-post linked decision making. One of the first factors is the translation of the environmental plan, the initial policy, into the municipality's own sectoral reference frameworks, i.e. the municipality's own (sectoral) strategic policy. It is not the environmental plan that is considered relevant by the respondents to their own decision making, it is their own memorandum or their own plan. A second factor appears to be the influence of government policy within certain policy sectors on municipal policies. Here the co-ordination between the policy sector in question and the 'environment' took place already at the government level. Rather than the environmental plan, sectoral state policy is seen as relevant by the respondents. A third factor could be regional influence. Policies are determined regionally, and regional policy is seen as being relevant rather than the environmental plan. A fourth factor that limits the relevance of the plan to ex-post linked policies is the newness of the policy. Previous to the environmental plan, other policy sectors have sometimes drawn up their own memoranda or plans, in which attention is paid to environmental aspects. Due to the influence on the planning process of the actors addressed, points of action from such sectoral memoranda find their way into the environmental plan.

Even if the environmental plan is considered relevant, this does not have to result in consultation of the plan. Various mechanisms can impede actual consultation. Environmental arguments can be introduced in consultation situations between the environmental department and the other departments. Certain decisions are weighed in full by this department itself.

Knowledge, relevance and consultation of a plan do not yet have to result in the use of a plan, in the sense that this plan influences considerations during ex-post linked decision making. A number of factors can reduce the influence of the environmental plan in the argumentation to zero. Here a clear distinction has to be made between the influence of the environmental plan and the influence of 'environment' in the argumentation in general. One of the first factors which could result in the plan not being used, although it is known, is considered relevant and was consulted, is the relation between policy sectors. A second factor is the politicisation of the policy issue. A third factor is finance. Business-economic freedom of choice is not always present.

13.7   Planning Form and Performance

The third of the questions addressed in this research is: To what extent does the influence depend on the form of planning? All the case municipalities were classified on the basis of their internal function: the extent of directionality. Their directional function, even for the most directional plans of the case municipalities, remains a limited one. All the plans were also aimed at plan realisation, to a greater or lesser extent. This hybrid character results in plans which pursue both directionality and plan realisation.

No clear relation was found either in the qualitative or in the quantitative part of the effectiveness study, between the presence of the performance pattern and the extent of directionality of the planning form. Even in the most directional plans, the intended performance pattern (knowledge of the plan, considering the plan relevant, consultation of the plan and the eventual use of the plan by the planning subject in the sense of referring to the plan in decision making) appeared to be present only in a small part of ex-post linked decision making.

A first explanation for the limited impact is the *character and function* of the plans. On the one hand, municipal environmental plans appear to have a mixed strategic and operational character, on the other hand they appear to have a hybrid character in terms of their function, in the sense that they are supposed to have both a directional function as well as being aimed at plan realisation.

A second explanation can lie in the *theoretical models* of the intermediary processes. The theoretical models used of the process between plan and ex-post linked decision are of a relatively simple nature. We have good reasons for limiting ourselves to simple models. To draw up rival theoretical models a relatively simple basic model is required. Otherwise it is virtually impossible to give the rival theoretical models an equal chance of being rejected. A more detailed model, which also looks at a classification of plan statements, for instance, may yield a more positive opinion on the effectiveness of directional framework plans.

Apart from using relatively simple theoretical models, we have also imposed on ourselves the limitation of a one-actor model. The argument for doing so is the supposed 'unity of planning and acting' in the decision-oriented planning approach. In our opinion this is an essential feature of the 'decision-oriented' performance study. If we abandon the unity of planning and acting, we arrive at a completely different form of 'performance research'. Performance research that is in line with the decision-oriented planning approach assumes that 'unity of planning and acting' is required. The plan-as-a-directional-framework directs ex-post linked decision-making. This presupposes a connection between plans and operational decisions. It also implies that many plan drafters and ex-post linked decision-makers somehow belong together; that they are part of a greater whole. The obvious criticism is that 'unity of planning and acting' is seldom found in government organisations, and that there governments are subdivided into various departments resulting in communication problems between decision makers who should formally be seen as part of the planning subject, is a very major one regarding environmental policy plans due to the integration nature of environmental policies and the large number of departments involved.

In the environmental policy plans of municipalities we have assumed this unity, because we have defined the mayor and aldermen as a planning subject. We do not distinguish between formal decisions taken by mayor and aldermen and official memoranda. Unity of planning and acting is assumed by the direct link between the drafter of plan (mayor and aldermen), the institution which approves the plan (council) and the ex-post linked decision-makers (mayor and aldermen or persons authorised by them). Justifying decrees are still the task of the mayor and aldermen, even although the preparation of such a decree is not a monolithic whole. 'Unity of planning and acting' remains valid also in the theoretical models. Here we do not use models involving more than one actor. The empirical findings of our study indicate that it is sometimes doubtful whether the planning subject can be seen as a monolithic entity. Knowledge and relevance of the plan are not always self-evident in other municipal departments.

However, this does not detract from the principle that a plan-as-a-directional-framework can only function as a reference framework for later decision making if there is 'unity of planning and acting'. After all, this reference framework is seen as an advance investment to be able to take 'better' decisions later on. However it may be that a plan-as-a-directional-framework does influence the ex-post linked decision making by other actors. The drafter of the plan and the ex-post linked decision maker do not have to be the same actor as long as there is a certain level of unity of planning and acting. Otherwise the plan cannot be of any significance to the quality of the ex-post linked decision making. In our opinion, work needs to be done on the 'unity of planning and acting' in environmental policy to enable environmental plans to really function as directional frameworks.

Thirdly we looked to the impact on decision making and not into other functions of plans like acquiring 'commitment' or political support, which could be called 'latent functions of planning'. As one of these latent functions of environmental policy planning we can mention the fact that an environmental plan can strengthen the position of the environmental aspect with respect to other interests. An evaluation which uses this latent function as a criterion could result in a far more positive opinion on environmental policy planning. Similar findings in performance research on the (in)effectiveness of plans have also resulted in further examination of these latent functions. In the present study we focus on the processes between the formal plan and ex-post linked decision, instead of on the influence of the way in which the plan was accomplished. This would require not just a different form of decision making, but also pass over the policy theory behind the planning framework. After all, this theory does *not* assume that desk-drawer plans have to be drawn up if necessary, because the effect on later decision making depends on a positive implementation process. In other words: we focus on the effects of the plan on ex-post linked decision making instead of the effects of the implementation process on ex-post linked decision making. We will deal with this in more detail below.

A fourth explanation concerning our conclusions has to do with research design choices. Due to the choice of population, part of ex-post linked decision making was intentionally excluded from the decision making population. Particularly those decisions that were mandated in part of the case municipalities – the most extreme form of autonomous official decision making – were left out of this population. This causes a distortion within the population of decision making that we studied, in the sense that it contains fewer operational decisions and operational decisions of a certain type because of this.

The question is whether this leads to underestimation of the value of the plan. The distinguishing criterion in the decision-oriented planning approach between operational and strategic decisions is the level of 'commitment'; the extent to which the planning subject commits himself. Our study has shown that environmental policy plans result in few decisions with a high level of commitment. As mentioned before, much ex-post linked decision making is of a strategic nature. Because of the gradual performance of initial environmental plans in ex-post linked plans and eventual day-to-day decision making, we obtain little information about the relation with eventual operational decision making.

13.8   Conclusions

As we indicated in section four environmental plans were used mainly when drawing up sectoral strategic plans and memoranda in relation to the plan, and when realising concrete points of action. This conclusion that environmental plans are ineffective in terms of performance results could be revised if, as some suggest, we take in to account a broader interpretation of the concept 'use of a plan', or expand the frame of reference for decision making outside the scope of the plan.

In the first of these situations, the concept 'use of a plan' can be more widely interpreted if we look beyond the instrumental 'use of a plan' also to conceptual and persuasive use. Our study shows that in municipal plans all three of these uses play a part. In the case of instrumental use in the sense of references to the plan, legitimising considerations may play a part. Some respondents explicitly indicated that they did use the plan in a conceptual way. Use of the plan was particularly seen when sectoral strategic plans and memoranda were drafted in relation to the plan and when concrete points of action were realised. In our opinion, to be able to speak of directionality of a plan, there has to be an instrumental use. Performance means clarification of decisions on the basis of the plan, which is more than just a conceptual using of the plan (opposing De Lange, 1995).

Concerning the second of these points, certain matters outside the scope of the approved plan may function as a frame of reference for decision making (Wallagh, 1994). Certain parallels can be drawn between the performance of municipal structural plans and municipal environmental policy plans. Structural plans are also strategic, indicative plans. Quantitative and qualitative performance research was performed into the structural plans of Amsterdam in particular. The quantitative performance research showed that in structural plans few written references to the plan are found in Municipal Council decisions (Wallagh, 1994). For structural plans this led to the conclusion that we should go in search of the 'world behind the plan'. Follow-up studies investigated how the various actors (the planning subject, higher authorities and private actors) dealt with the structural plan in concrete operational decisions. The conclusion with regard to the municipal actors reads:

> The various actors are acquainted with the plan. Its main outlines are subliminally present and have a performance on behaviour. In day-to-day decision making the plan is not used as a manual. Rather it functions as a reference book from the actor's own, often specific background. For concrete objectives the municipality often resorts to its own sectoral plans. (Wallagh, 1994, p. 264)

This conclusion, which is in line with conclusions from previous performance studies, shows clear similarities as well as clear differences with conclusions from our study. Also in our study we have, in a sense, opened up the 'world behind the plan'. We also looked at both the implementation process and the use of the plan, focusing on the use process. Our study showed that in the case municipalities where according to the respondents 'it was not permitted to be absent from an environmental policy plan meeting', knowledge of the plan, its relevance, consultation and use of the plan were highest, relatively speaking. Here we agree with the conclusion that the planning process has a major influence on the use of the plan. However, we object to the concept of 'invisible results', since this might imply that we would have to satisfy ourselves with the existence of a desk-drawer plan, because the implementation of such a plan would have yielded positive invisible results such as agreement, commitment and more intensive contacts. We called these results 'latent functions of planning', i.e. the individual value that a planning process may have. We feel that here we should be careful not to discard an essential feature of a 'plan-as-a-directional-framework'. This essential feature is that to establish links with other decisions, a frame of reference should be used: a plan. The plan functions as an aid to ex-post linked decision making.

Improvement of the Influence of Environmental Policy Plans

Organisational solution. Firstly the decision-oriented planning approach itself offers a number of solutions to the desired relation with the planning process. The inter-organisational planning suggestions which were made within the strategic choice approach were not adopted in the operationalisation of the planning conception within the Dutch planning system. One of the lessons which may be drawn from the utilisation study of social-scientific research, on which the extension of the use concept was based, concerns the importance of key actors. For the performance of environmental plans the presence of such a key actor within other departments, who propagates the plan within his own department, appears to be of great importance. Concerning the inter-organisational performance of the strategic choice-approach, uncertainty appears to exist about the activities of related policy sectors (class UE). Much of this uncertainty originates from other organisations that operate within the policy field of the planning subject. Friend et al., (1974) use the concepts 'reticulist' and 'connective planning'. The development of a planning process depends partly on the capacity of key actors to build informal networks of personal contacts. Persons with these capacities are called 'reticulist'. Friend et al. (1974) advocate the placement of such reticulists in strategic positions as an important condition of inter-organisational planning.

In other words, a significant form of uncertainty concerns the question of what other departments intend to do with the plan. Precisely in this aspect the decision-oriented planning approach appears to rival blueprint planning. Blueprint planning assumes that conformity of other departments to the plan is a matter of applying enough instruments of power. The decision-oriented planning approach sees this as an uncertainty problem.

Here we should also realise, however, that it is of great importance what the advance intentions of the planning subject were with his planning. To put it very strongly: those who pursue a final picture will fail to take into account the problems caused by the involvement of third parties (other departments) in the planning process. Vice versa, the blueprint will also be more strongly oriented toward the department drawing up the plan. In determining the extent of target achievement we found that for the two municipalities furthest to the right on the directionality continuum, i.e. most similar to the planning form 'blueprint planning' showed a significant decline in the number of initiatives from other services, while the other municipalities did not show any significant changes.

Our empirical findings showed that the 'unity of planning and acting' is partly determined by factors such as political emphasis on the importance of the plan, and is likely to vary between departments. 'Unity of planning and acting' at the municipal level is not self-evident, but it is necessary in order to realise the directionality of plans. Work will have to be done in environmental policy on this 'unity of planning and acting' if environmental plans are to really function as a directional framework. The actors within the planning subject should pursue the creation of a joint 'plan-as-a-frame-of-reference'.

Key actors or reticulists appear of crucial importance to the use of plans in such a context. This also means a shift in the perception of uncertainty from uncertainty class UR – uncertainties with regard to choices or decisions which can be taken in future by the subject itself – to uncertainty on the activities of related policy sectors (class UE).

In the case municipalities environmental policy plans were not always drawn up as a joint frame of reference. The involvement of the different actors within the planning subject in planning varied a great deal between the case municipalities we studied. Some plans have a strong focus on the service which drafted up the plan. They function as a plan in which an opinion is given on developments which are desirable from the point of view of the environment. In environmental terms we could speak of plans where the 'leapfrogging' with other policy sectors has yet to begin. First a translation has to be made into the strategic sectoral plans of the municipality itself. Then there can be a performance at the strategic level. The question than is what we will find at the sectoral operational level of the performance.

Solutions in the Content of the Plan

The planning conception itself indicates an important reason why plans are insufficiently able to function as directional frameworks. Plan statements should respond explicitly to any (operational) decisions for which they intend to be a framework (Mastop, 1984, p. 274). This was noted also in policy practice. The Central Council for Environmental Hygiene (CRMH) states in its comment on the memorandum: 'More than the sum of its parts' that 'a planning statement should be capable of having an impact' (1985).

The previous paragraphs show that environmental plans are insufficiently able to function as directional frameworks, because the plans are too hybrid in nature, i.e. they are too much split in their intentions. They aim both to realise points of action and to play a part in the argumentation of ex-post linked decision making. Summing up concrete actions appear to be of major importance.

In themselves, concrete actions may also result in a performance on ex-post linked decision making. However, this is not the most interesting part of performance, and besides this part is relatively easy to accomplish. The problem is, though, that the points of action were formulated on the basis of a 'final picture' concept; realising points of action drawn up in advance. If there is to be a performance, statements should be formulated in such a way as to be able to influence the choice situation of the ex-post linked decision maker. Plan statements should be aimed at directionality instead of at plan realisation.

An alternative appears to be using so-called degrees of hardness. A planning statement indicates when an ex-post linked decision is so essential that the planning subject has to take the reference framework of the plan explicitly into account. This suggestion is in fact made within the planning system. However, in our study we found only one example of such a construction, in traffic.

A second problem in strategic environmental policy plans lies in the nature of municipal environmental policy. It is typical of environmental policy that much of environmental policy needs to be realised through other policy fields. Due to the importance of external integration we find a certain type of ex-post linked decision making. Much of this ex-post linked decision making is, as we said, of a strategic nature. Due to the gradual performance of initial environmental plans in ex-post linked plans and memoranda, the direct relation to operational day-to-day decision making is limited.

Much of environmental policy is not, or is hardly, developmental policy, rather it is built around limiting instruments to counteract threatening developments (Salet, 1994). Therefore, environmental objectives may clash with the overall strategic objectives of municipalities. This aspect was clearly seen in the case municipalities due to the difference between municipalities focused on economic growth (and therefore employment) and the municipalities with a 'green' orientation.

The question is whether an environmental plan can function as a directional framework if it contains mainly 'protective' policies. If the environment is to play a real part in the developmental policy of a municipality, it should either be given a central place in city development plans, or the environmental plan should be seen as being equivalent to city development plans. In analogy to the provinces, municipalities may in future have to draft integral plans for the surrounding environment.

Notes

1   Dr. F.H.J.M. Coenen works at the Faculty of Public Administration and at the 'Centre of Clean Technology and Environmental Policy' (CCTEP), both of the University of Twente. Dr. Coenen holds a masters degree in Public Administration and a PhD in environmental policy planning. He publishes and lectures in the field of planning, environmental policy and evaluation research.

2   In three municipalities, the percentage of external integration decisions rises from period 1 to period 2, but this increase was only significant in two. In two others we found a significant drop from period 1 to period 2. From period 2 to period 3 all municipalities show an increase. However, this increase is significant only in two municipalities. No significant rise is seen in the initiatives from the other services.

References

Alexander, E.R., and A. Faludi (1989) Planning and Plan Implementation: Notes on Evaluation Criteria, *Environment and Planning B: Planning and Design*, Vol. 16.

Ashworth, G. (1992) *The Role of Local Government in Environmental Protection: First line defence*, Longman, Harlow.

Ashworth, G.J., and P.T. Kivell (eds) (1989) *Land, Water en Sky, European Environmental Planning*, GeoPers, Groningen, The Netherlands.

Barret, S., and C. Fudge (eds) (1981) *Policy and Action: Essays on the Implementation of Public Policy*, Methuen, London.

Blowers, A. (ed.) (1993) *Planning for a Sustainable Environment*, a report by the Town and Country Planning Association, Earthscan, London.

Coenen, F.H.J.M. (1992) *The Role of Municipal Environmental Policy Plans in Environmental Management*, paper XXII International Congress of Administrative Sciences, Vienna.

Coenen, F.H.J.M. (1996) *The Effectiveness of Local Environmental Policy Planning*, CSTM

Studies and Reports No. 2, Enschede, The Netherlands.

Faludi, A. (1980) Implementation *Planning or the Implementation of Plans? Effectiveness as a Methodological Problem*, Werkstukken van het Planologisch en Demografisch Instituut, No. 31, Amsterdam.

Faludi, A. (1987) *A Decision-Centred View of Environmental Planning*, Pergamon Press, Oxford.

Faludi, A. (1988) Two *Forms of Evaluation*, Werkstukken Planologisch en Demografisch Instituut, Amsterdam.

Faludi, A. (1989) Conformance vs Performance: Implications for Evaluations, *Impact Assessment Bulletin*, Vol. 7.

Faludi, A., and J.M. Mastop (1982) The I.O.R. School: The Development of a Planning Methodology, *Environment and Planning B: Planning and Design*, Vol. 9.

Friend, J.K., and W.H. Jessop (1969, 2nd ed. 1977) *Local Government and Strategic Choice: an Operational Research Program to the Processes of Public Planning*, Pergamon Press, Oxford.

Friend, J.K., J.M. Power and C.J.L. Yewlett (1974) *Public Planning: the Inter-Corporate Dimension*, Tavistock Publications, London.

Hall, D., M. Hebbert and H. Lusser (1993) The Planning Background, in A. Blowers (ed.), *Planning for a Sustainable Environment*, a report by the Town and Country Planning Association, London.

IFHP/IULA (1984) *Changing Roles for Local and Regional Governement in Environmental Management*- Extra *Burdens or New Opportunities*? Procedeedings International Symposium, Maastricht, The Netherlands, December, VNU, Science Press, Utrecht.

Lange, M. de (1995) *Besluitvorming Rond Strategisch Ruimtelijk Beleid*. Verkenning en Toepassing van Doorwerking als Beleidswetenschappelijk Begrip, Thesis Publishers, Amsterdam.

Mastop, J.M. (1984) *Besluitvorming, Handelen en Normeren*, PhD-Thesis, Amsterdam.

Sustainable Cities Project (1994) First report October 1994, EU Expert Group on the Urban Environment.

Salet, W.G.M. (1994) Institutionele voorwaarden voor stedelijke milieupolitiek, *Beleid en Maatschappij*, No. 4.

Thomas, H.D., J.M. Minett, S. Hopkins, S.L. Hamnett, A. Faludi and D. Barrell. (1983) *Flexibility and Commitment in Planning*, Nijhoff, The Hague.

Wallagh, G.J. (1994) *Oog voor het onzichtbare: 50 jaar structuurplanning in Amsterdam 1955-2005*, PhD-Thesis, Amsterdam.

Ward, S. (1993) Thinking Global, Acting Local? British Local Authorities and Their Environmental Plans, *Environmental Politics*, Vol. 2, No. 3.

Weiss, C.H. (1977) Increasing the Likelihood of Influencing Decisions, in L. Rutman, *Evaluation Research Methods: a Basic Guide*, Sage, Beverly Hills.

Wildavsky, A (1973) If Planning is Everything, Maybe it's Nothing?, *Policy Sciences*, Vol. 4, No. 2.

Part D
Focusing on Integration at the
Neighbourhood Level

Part D
Focussing on Integration at the
Neighbourhood Level

Chapter 14

Ecological Renewal of Post-war Neighbourhoods: Problems and Solutions

A. Bus[1]

14.1 Introduction

Most human activities take place within urban areas. These activities involve flows of materials and goods supporting both urban economic activities and human consumption. Environmental problems such as resource depletion, waste production and soil contamination are clear signs that these flows have grown out of control. This problem can not be reduced solely to an economic, social, psychological or physical planning problem. However a disciplinary approach such as urban planning may contribute to a better understanding of the ways cities can become more 'environmentally friendly'. Urban planning can in this case be denoted as 'ecological urban planning'. Ecological urban planning is far from simple. One reason is the complex, heterogeneousness of urban structure. To reduce the complexity of the urban structure this chapter focuses on neighbourhoods.

So far, most attention has been given to applying principles of ecological planning in the design of new neighbourhoods. However, the vast majority of households live and work in already existing urban areas. For example, the Dutch government has estimated that for the period from 1995 up to 2015, 1,312 million dwellings must be built (VROM, 1992a). In 1995 the housing stock in The Netherlands consisted of 6,285 million dwellings (VROM, 1996). This means, that in 1995, 80 per cent of the Dutch housing stock of 2015 already exists. This is an important signal for paying more attention to the implementation of ecological measures for processes that are traditionally described as 'neighbourhood renewal'. In this chapter this approach is called 'ecological neighbourhood renewal' or more briefly, ecological renewal. Ecological renewal is defined as the implementation of recuperative changes in the physical neighbourhood, in response to pressures from economic, social and physical problems, that reduce or prevent environmental degradation.

In the past little attention was paid to the environmental aspects that are so important today during the design of neighbourhoods. Therefore it may be assumed that the environmental quality of existing neighbourhoods can be improved by renewal plans. Unfortunately, urban renewal processes in Europe have mainly been guided by socio-economic considerations (Alterman and Cars, 1991).

The housing shortage after the Second World War in the Netherlands led to an enormous production of new houses, which lasted well into the 1980s. During four decades, some 100,000 new homes were built every year. In the years immediately after the War, 'quantity' was the leading principle of Dutch housing policies. The vast majority were subsidised by the state, which enabled rents to remain low and the building process to continue. In the decades to follow, however, the emphasis slowly shifted to 'quality'. Many post war neighbourhoods, especially those that were built in the fifties and sixties became objects for renewal projects. Nowadays, neighbourhood renewal is focused on post war areas (Agricola, 1995; Drontman, 1996).

14.2   Post War Neighbourhood Renewal as a Process

Ecological renewal of post war neighbourhoods should for practical reasons match the more traditional approaches to neighbourhood renewal. Urban renewal can be described as the form of recuperative change in the physical city by which outworn or outmoded structures and facilities, and in time whole areas are altered or replaced in response to pressures of economic and social change. In this sense, urban renewal is a process that has been going on as long as cities have existed and flourished although the reasons behind urban renewal have changed (see Priemus and Metselaar, 1992). To make sure that besides solving the existing problems, neighbourhoods could function for a period of at least 20 to 25 years without the need for big and costly improvements, planners anticipate trends and developments including demographic changes, technical improvements, political engagement and economical developments. An important lesson learned from neighbourhood renewal projects is that every neighbourhood has its own problems and possibilities and that it is very important for local authorities to involve other institutions and inhabitants in an earlier stage in neighbourhood renewal processes (Tjallingii, 1995).

In Figure 14.1 the main relations that influence ecological neighbourhood renewal are visualised. Ecological renewal activities will only be accepted if they provide solutions for existing neighbourhood problems. In addition to the traditionally addressed socio-economic and physical problems, neighbourhood renewal plans have now also the responsibility to prevent or reduce deterioration of the environmental quality. In other words, ecological neighbourhood renewal requires that environmental protection has a full place in neighbourhood renewal plans.

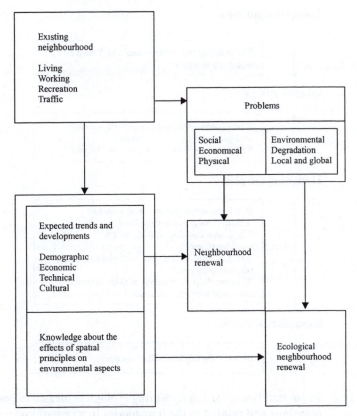

**Figure 14.1   Ecological neighbourhood renewal: differences between a
traditional and an ecological approach**

14.3   A Method to Implement Ecological Measures in Post War Neighbourhoods

Although in practice ecological renewal of neighbourhoods seldom takes place, ecological neighbourhood renewal can cover about 1.4 per cent of the total housing stock on an annual basis (Bus, 1996). The challenge is to place the ecological content smoothly within a neighbourhood renewal process. In broad lines a neighbourhood renewal process can be divided into four phases, taking the initiative, integral analysis, designing the plan and execution of the plan (Bus and Heins, 1997). During these four phases there are eight issues which are very important for implementing ecological measures. These issues can each be posed as a question. These questions are stated and illustrated below but first a short overview of the relationship between the four phases and the posed questions will be presented in Figure 14.2.

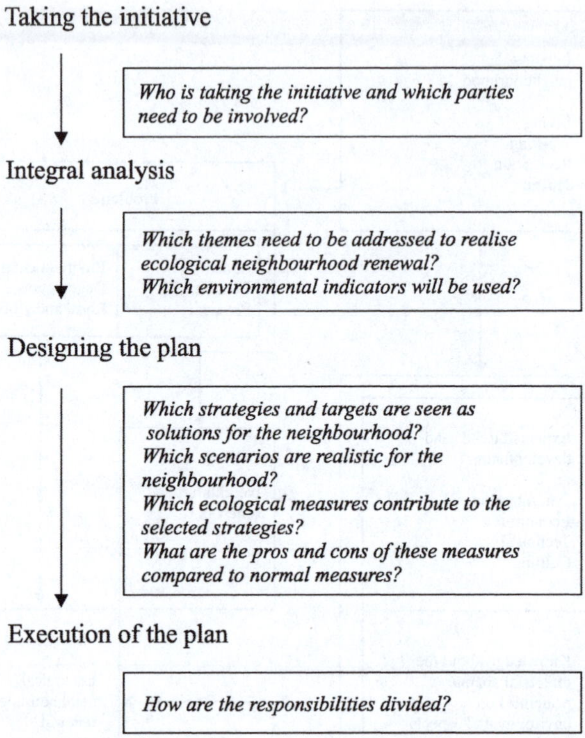

Taking the initiative

> Who is taking the initiative and which parties
> need to be involved?

Integral analysis

> Which themes need to be addressed to realise
> ecological neighbourhood renewal?
> Which environmental indicators will be used?

Designing the plan

> Which strategies and targets are seen as
> solutions for the neighbourhood?
> Which scenarios are realistic for the
> neighbourhood?
> Which ecological measures contribute to the
> selected strategies?
> What are the pros and cons of these measures
> compared to normal measures?

Execution of the plan

> How are the responsibilities divided?

**Figure 14.2   Important issues for implementing ecological measures posed in
questions and related to the four phases in a renewal process**

Taking the Initiative

During the 1960s and 1970s, spatial planning switched from blueprint planning towards process planning. According to Habermas' planning, a communication process is needed to obtain maximum support for decisions. The recognition grew that planning is a social process (Voogd, 1995). The generation of ideas is considered to be a result of social forces and not (only) the creative powers of designers. It has become obvious that the participation of all the actors is needed in order to create sufficient public support for all the measures and projects from which is expected that they will improve the quality of the neighbourhood. Especially if more drastic solutions, like the reduction of social housing are promoted. The actors in the planning process for neighbourhood renewal will vary with the kind of community and the problems faced. Although the local government will always be an actor, because of its responsibility to maintain liveable neighbourhoods, the question arises whether the local government should always be the main actor in taking the initiative in neighbourhood renewal. The actor that has a main interest in renewal activities of a certain neighbourhood could take the initiative for an integral planning process as well. With respect to ecological neighbourhood renewal, it is also important to know which party takes the responsibility to put the environmental quality on the local agenda.

Integral Analysis

Neighbourhood renewal seldom takes place solely for environmental reasons. This does not mean that we can not pay attention to ecological aspects when neighbourhood renewal takes place. However if we want to improve the environmental quality of a neighbourhood it is necessary to realise that ecological friendly measures which can improve the environmental quality need to fit in a broader context. However, before the existing problems and expected threats for any neighbourhood can be countered, it is necessary to determine which processes are responsible for these difficulties. To find all the processes that matter, a broad analysis is necessary. This broad analysis can only be successful if all the parties involved use their resources, knowledge and data.

In most neighbourhoods, existing problems refer to social problems, so it is obvious that social aspects should be part of this analysis. For corporations, house owners and shopkeepers it is important to know the position of their housing and business estates in the context of the regional housing market. Another aspect concerns the pros and cons of the spatial structures and patterns of the built and non-built areas. Of course, to be able to determine if neighbourhood renewal measures will improve the environmental quality it is first necessary to determine the existing environmental quality. Apart from the description and mapping of all the existing problems and expected threats, it is also essential to give a description of the strengths and opportunities in the fields mentioned before.

To measure the environmental quality of a neighbourhood it is necessary to have or to develop a range of features that are measurable and comparable to a set of 'standards', which in turn are based on generally recognised criteria of health, safety, waste management and other elements of public and ecological interest. For example standards relating to the design of lots, blocks, streets, population densities, and the type and intensity of development are included in subdivision regulations and zoning ordinances. Adopted standards relating to building materials, occupancy, and sanitary facilities often are specified in building, housing, fire and sanitation codes. Once objective measurable environmental indicators are defined a diagnosis can be made with relative ease and changes measured.

Designing the Plan

Dependant on the outcome of the analysis, the parties involved can decide which maintenance measures and development projects are necessary to revitalise the neighbourhood. Measures in the field of maintenance or often small, and aimed to improve the safety, to decrease the criminality, or to improve the structures in a neighbourhood. Development projects often have a large physical impact that require long-term investments; for example, create more differentiation in the existing housing stock, improve the spatial quality and reduce environmental degradation (Bus and

Heins, 1997). However, the parties involved in a neighbourhood renewal process often have different opinions about what is good or wrong for their neighbourhood. Therefore, it is essential to investigate the various opinions of these parties. At this point, reaching consensus concerning the set of strategies which are desirable for the neighbourhood is a very important first step. During the design of these strategic goals, it is important to involve also the actors responsible for working out these strategies in detailed programs and the execution of these programs. At the end all the differently oriented measures and projects, necessary to improve the neighbourhood, need to be integrated in one strategic plan. Intensive communication between designers, managers, dwellers and working people is therefore essential.

To make it possible to let a neighbourhood function well over a period of at least 25 years without the need for new costly improvements, it is wise to make use of local scenarios. With the help of these scenarios local authorities can reach their decisions which improve the chance on realising the implementation of a neighbourhood renewal plan. During the design of local scenarios the current situation is of course the point of departure. Together with anticipated external trends and developments, and policy targets at a national and regional level, legislation sets up the framework that determines the playing field of the local authorities and other groups involved. Local scenarios are in this step not seen as blue prints for utopian futures but as possible directions where one or two developments are stressed. For example, a local scenario in which people spend more time at their homes by tele-working and where recreation near the house is very important. Or a scenario where mobility becomes even more important and the demand for car facilities becomes even bigger.

The next challenge is to find ecological measures which offer a contribution to the formulated strategies and fit with the designed local scenarios. From a spatial planning point of view three principal solution directions can be taken to reduce environmental degradation at the level of a neighbourhood. Changes in the design of spatial structures, e.g. road lay-out or public green, improvements to the existing housing stock such as small scale renovations, demolition and rebuilding houses, and neighbourhood management. Obviously, it is not possible to consider within these three main directions all possible solutions that can reduce environmental degradation. For practical reasons a reduction in the number of possibilities is necessary. This is possible by dividing spatial planning solutions into three categories: solutions which are already implemented in neighbourhood renewal projects or elsewhere, solutions that are already invented but not implemented, and solutions that are completely new. Analysing these three categories with the three main directions offered by spatial planning, nine possible fields appear (Table 14.1). In practice it is most likely that actors involved in neighbourhood renewal processes only will consider solutions that are already implemented with success in other existing neighbourhoods.

The final task is to reach consensus about the renewal plan that will be implemented. In this context it is important to visualise, for the involved parties, the pro and cons of the alternative measures on their environmental effects but also in a broader scope; on their financial as well as non-financial consequences. In this respect ex-ante multi-criteria evaluation methods might be helpful.

Table 14.1   Categories of spatial planning solutions

Spatial Planning Solutions	Implemented	Invented but not implemented	New
Spatial design	I	II	III
Housing stock	IV	V	VI
Management	VII	VIII	IX

Execution of the Plan

Too many plans are laying in drawers, because they lack the support of one of the involved parties: a housing corporation has other ideas; action groups of tenants are against any measures that will raise their rent; or shopkeepers are frustrated while they were not involved in the planning process. A top-down approach has in other words its limitations. To tackle these disadvantages it is often suggested that an integrated bottom-up approach is necessary as well as reaching consensus between the various participating actors.

14.4   Implementation Problems

Given the large potential for reducing environmental degradation through ecological neighbourhood renewal (Bus, 1996) and knowing that in practice ecological renewal of neighbourhoods seldom takes place, this section pays attention gives a short overview of the problems one can expect in the different phases already mentioned in paragraph four. As we know, without an explicit, positive concern for environmental aspects, little improvement on environmental quality can be expected. It is in other words a problem of the first order if environmental quality not is mentioned as an important theme in the coming neighbourhood renewal process. The implementation problems of ecological measures occurring in neighbourhood renewal processes were explicit attention is paid to the environmental quality can in this respect be seen as problems of the second order. Vinkhuizen a neighbourhood in the city of Groningen were explicit attention is paid to the environmental quality will be used as a case to point out some second order problems which can occur during the II, III and VI phase of a neighbourhood renewal processes. Before providing an overview of these problems, it is useful to briefly describe Vinkhuizen and explain the reason for taking actions in this neighbourhood.

Vinkhuizen is a neighbourhood in the north-west of Groningen, built in the post-war period between 1965 to 1975, with about 11,300 inhabitants. Vinkhuizen is on its east side more or less separated from the rest of the city by the city ring road and a channel called *Reitdiep*. The northern part of Vinkhuizen borders on countryside. The western part of Vinkhuizen borders on a park. In the south, Vinkhuizen borders on sport fields and an industrial site called *Hoendiep* (Figure 14.3). The neighbourhood is divided into six parts. The classification of these parts is based on geographic location, types of dwelling, and ownership proportion (Figure 14.4). The following is based on the south-west part of neighbourhood Vinkhuizen.

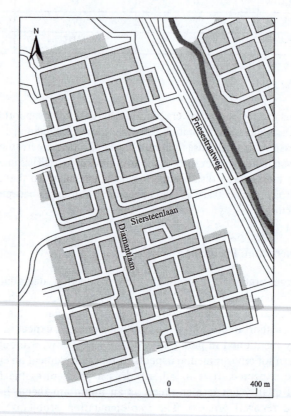

Figure 14.3   Neighbourhood Vinkhuizen
Source: Buro Vijn (1996)

A housing corporation, '*the Huismeesters*', owns about 1000 dwellings (apartments, elderly houses and one family houses) in part 6 of neighbourhood Vinkhuizen and is concerned about the rentability of their dwellings. Although at this moment there is no structural vacancy of their dwellings the expectation is that it will become a problem in the near future. There is a slight increase of social problems while at the same time increasing demand for private dwellings. The corporation is afraid to loose residents with higher incomes to a new neighbourhood, *De Held*, planned to be built next to Vinkhuizen. Therefore, the corporation wants to improve, in close corporation with their residents and the municipality of Groningen, the quality of their housing stock and the public space in the south-west part of neighbourhood Vinkhuizen. By improving the quality of the neighbourhood it is to be expected that the neighbourhood can fulfil its functions for a longer period of time without any need for costly structural changes (see section 14.2). This will not only improve the position of the corporation, their residents and the municipality, but it will also contribute to a more sustainable society.

Taking the Initiative

A problem with respect to this phase is that with the introduction of a new policy for urban renewal (VROM, 1992) the rules were changed. The national government distanced itself from its financial and intrinsic responsibilities with respect to public housing. This raised a lot of questions by the local parties about how to deal with problems in their neighbourhoods. All the parties involved see that it is important to undertake action but it is not clear who has which responsibilities. In the search for their own role in this new situation, complex matters such as integrating environmental measures in renewal plans have received little attention.

An inventory by Ketelaar et al. (1997) pointed out that in only six of the twenty seven largest cities in The Netherlands environmental aspects as seen as important enough to give it an explicit place at the beginning of a neighbourhood renewal process. Research results from Rigo (1996) underlined these findings as well as a survey made by the Dutch Foundation Nature and Environment. The reasons mentioned by these organisation for not paying explicit attention to environmental aspects in neighbourhood renewal processes were, besides the lack of clarity about responsibilities, the fact that, in contrast to economic interests (shop owners, building companies etc) and social interests (dwellers, police etc), environmental interests lack natural representatives on the neighbourhood level. The Dutch government required municipalities to finish their task to build new houses in the so called 'VINEX-locations'. Vinex-locations are locations designated by the Ministry of Housing, Spatial Planning and Environment of the national government of The Netherlands (VROM, 1993). The result of this was that municipalities who are environmental minded focus their time and resources on these new development plans rather than on renewal plan. Another reason was lack of knowledge. Lack of knowledge of how to integrate environmental aspects in a complex process as neighbourhood renewal and lack of knowledge concerning different intrinsic aspects. In order to enable a fruitful exchange of opinions between persons and institutions about the concepts of living area quality and sustainability, it is essential to elaborate these concepts in the framework of a neighbourhood renewal process. In practice such a framework is not available. Probably due to the fuzzy character of these two concepts it seems difficult even for municipalities to develop such a framework.

In the case of Vinkhuizen the following framework is accepted by all the parties involved. From an urban planning approach, the quality of a neighbourhood is determined by aspects that influence the housing stock quality and the residential area quality. Aspects that influence the quality of the housing stock include technical aspects, maintenance, and arrangement of houses. Aspects that determine the quality of the residential area include existing facilities, the quality of facilities, accessibility to facilities, identity of a neighbourhood, public security, environmental quality, ecological quality (green structures), street litter and dog dirt.

The same aspects that determine the quality of a neighbourhood are also important in obtaining an idea about the sustainability of a neighbourhood. It is to be expected that neighbourhoods characterised by a 'high residential area quality' can fulfil their functions for a longer period of time. In this case the word 'sustainable' means a neighbourhood which can function well over a period of twenty-five years without any need for costly structural changes. Environmental quality is just one of the aspects that determines the quality of a neighbourhood.

Environmental degradation occurs both at a local level (liveability) and a global level (sustainability). These levels are closely related to each other. Environmental quality at the local level has a direct impact on the quality of life in cities, villages, and areas designed for nature and recreation. It is therefore important to prevent or reduce environmental hindrance caused by activities in a neighbourhood or its surroundings. Aspects that influence the local environmental quality of a neighbourhood include risk contours, contamination of soils, water pollution, air pollution and nuisance caused by noise, odour and dust. Degradation of the environmental quality at the global level has an indirect impact on the quality of a neighbourhood. In the longer run, environmental degradation at a global level will negatively affect the local environmental quality, and thus the living quality of that neighbourhood. The possession of a garden is for example less enjoyable if the ozone layer is destroyed because this makes sun bathing a dangerous activity. It is therefor also important to prevent or to reduce the contribution of a neighbourhood to environmental degradation at the global level. Aspects that influence the global environmental quality include preservation of biodiversity, ozone depletion, acidification, production of greenhouse gases, disposal of waste, squandering of resources, destruction of ecosystems, dehydration, eutrophication and diffusion of pollution.

Developing a masterplan for Vinkhuizen which would serve the goal of improving environmental quality and sustainability has revealed also a number of problems in the other phases. These problems, which likely affect similar efforts in other neighbourhoods, are identified and briefly discussed in the following paragraphs.

Integral Analysis

The local environmental quality in Vinkhuizen is evaluated on the following aspects: risk contours, soil contamination, water pollution, air pollution, noise nuisance, odour hindrance, dust hindrance and street litter. This was not very difficult, because the municipality of Groningen had data available for most of these local indicators, although it was mainly gathered at city level (municipality of Groningen, 1995). Data about water quality was available by the Province and information about street litter was obtained from inhabitants.

Formulating global environmental quality indicators was more difficult. This is partly due to the fact that it was hard to measure the current contribution of the south-west of Vinkhuizen to the environmental degradation at a global level in every detail.

Nevertheless, a realistic estimation of the current flows of materials and goods with respect to global environmental aspects (see Tjallingii, 1995) is necessary to obtain insight in the possibilities to reduce this degradation during the coming neighbourhood renewal process. South-west Vinkhuizen is analysed on its possibilities to reduce the use of water, fossil energy, materials, space for urbanisation, production of waste, and preservation or improvement of ecological areas.

One of the problems was that the data used to make a realistic estimation of the neighbourhoods current contribution to environmental degradation was often gathered for larger spatial units than the south-west of Vinkhuizen and for other purposes. Next to the dangers resulting from aggregation methods and different operational definitions the more general dangers in relation to validity and reliability need to be watched as well (see Etzioni and Lehman, 1967). Besides that, useful data was divided over different organisations and within an organisation often divided over many divisions and persons (see also Sawicki and Flynn, 1996).

Although for the case Vinkhuizen a few global environmental indicators are selected and even measured the main problem is that it is very hard to find or formulate standards to compare the results with. For example, reduction of space for urbanisation by building new houses in open places in existing neighbourhoods (often proposed by environmentalist in the western part of the Netherlands) will not succeed in a region were the supply of houses exceeds demands, as is the case in Groningen. Due to different local demographic, economic, technical and cultural circumstances each neighbourhood has other possibilities. A neighbourhood in Amsterdam will probably select other measures to improve the environmental quality than a neighbourhood in Groningen. It seems therefore impossible to insist that every neighbourhood reaches the same environmental standards or can implement the same solutions.

Designing the Plan

During the design of a renewal plan for a neighbourhood it is always difficult to foresee future developments that may be important for the neighbourhood. A good example in this case is the lack of adequate information about the developments in the regional housing market. The city of Groningen expected for example that from 1995 till 2000 their population would grow with 1000 inhabitants a year. Instead, the population shrunk annually with 1000 inhabitants a year. This is partly due to a national economic revival with the consequence that many students immediately after their graduate exam leave Groningen for a job elsewhere, partly due to a reduced numbers of new students and partly due to the fact that people with middle and higher incomes left the city to settle in one of the many villages near Groningen. Research pointed out that there are too many low-rent apartments (about 5000) and too few single-family houses, which can bind households with middle and high incomes in the city of Groningen. The housing corporations who own almost all the low-rent apartments in Groningen are aware of this situation, but each of them is not able to decide which of its low-rent apartments need to be renovated, or which demolished and replaced by more attractive houses to sell to middle and higher income groups. For

Vinkhuizen this means that as long as there is no city housing plan co-ordinated by the municipality and supported by all the corporations, these corporations are afraid to undertake action on a large scale.

To enlarge the chance that ecological measures will be realised in a neighbourhood renewal plan it is important to formulate a strategy which aims to improve the environmental quality. Although the local environmental quality in Vinkhuizen gave no reason to worry, all the parties agreed that more attention to the global environment was justified. Normally spoken a integral renewal plan should take care of this but because of the fact that the environmental quality was more or less neglected in existing neighbourhoods special attention was paid to this theme. In Vinkhuizen the following strategies were formulated: improve the commitment of the dwellers to their neighbourhood, improve safety, improve identity, enlarge the diversity of the neighbourhood and improve the environmental quality.

One of the problems Vinkhuizen met in the search for measures that can improve the environmental quality was the lack of knowledge about environmental cost-effectiveness. Currently, not much is known yet about the impacts and effectiveness of already implemented ecological principles elsewhere. This makes it difficult to persuade institutions and persons to implement ecological renewal principles. Knowing the costs of the different ecological renewal principles without knowing the effects and the effectiveness of these principles is not conducive to the implementation of these principles. Research aimed at obtaining more clarity about the costs, returns and other consequences is therefore of great importance.

One of the difficulties encountered during the search for more clarity is how to deal with different lifetime cycles of systems. For example, one of the goals is to reduce the total amount of fossil energy for heating for the whole neighbourhood. The idea of changing to another heating system received too much resistance from the energy company. Their investments in the existing system will become worthless. This option also received much resistance from the corporations and private owners because these had to invest in new heating equipment as well. This leaves only two other possibilities to reduce the use of fossil energy for heating. When the heating equipment needs to be replaced, only systems which deserve of best technical means can be chosen. Unfortunately, between 1992 and 1995 the corporation has replaced all their old heating equipment but not according to the standards of best technical means. Nevertheless there is still an improvement compared to the old situation. Another possibility is the use of alternative heating sources next to the existing heating system. This option is not seriously considered yet. This means that in the near future no further reduction in fossil energy use can be realised. At which time this will change depends on the quality of the existing equipment but also on the prices of energy and technical developments.

Another problem concerns the participation of the inhabitants not only during the initiative and analysis phase but also during the design and execution of the plan. Inhabitants and other parties involved often have different opinions about the quality of their living area, and the responsibility their neighbourhood has towards sustainability. Therefore, it is essential to investigate the various opinions of the parties involved including the inhabitants of the neighbourhood. After all, a neighbourhood renewal plan can only be implemented if it has enough support. In the case of Vinkhuizen the residents were not well organised. This has changed now they are supported by a professional organisation, which is paid by 'the Huismeesters'.

Closely related to the above mentioned point is the concern about the exchange of information between the involved parties. Before an appropriate selection of measures can be made, information is needed about many technical, social, environmental and economic aspects. This requires, in addition to a professional level of knowledge on the part of the different parties involved, a high degree of co-operation between these parties. Especially in the case of exchanging information concerning the environment it is important for those involved to understand its importance.

Execution of the Plan

An integrated bottom-up approach was recommended in the last paragraph of section 14.3. However in practice this bottom-up approach has proved to be quite difficult (Bus and Heins, 1997). One of the reasons for this is the difference between a maintenance and development approach. Preparations for the announced projects often consist of starting specific research programmes of selecting different alternatives within given conditions while many maintenance measures can be implemented immediately. For example, it is possible to insert an elevator in a three floor apartment within a few months, the benefit of this investment, however largely depends on whether these apartments will have a vital place in coming renewal plan or not. Although a good integration between these two approaches will arguably improve the quality of neighbourhood renewal plans, in practice they are often strictly separated from each other. A development plan is mostly designed before any attention is given to maintenance aspects. Furthermore, strategic plans and maintenance schedules are designed by different institutes and within these institutes different departments of persons with different goals. This sectoral approach presents itself especially in the execution phase, because in this phase the general measures laid down in a strategic masterplan need to be worked out in more detail including financial responsibilities. Many institution, departments and persons are then afraid to loose the control over their budgets and flee back to their sectoral roots. This is also partly due to the fact that these institutions, departments or persons are bounded to sectoral financial programmes which are established for several years and leave little room for changes. Another aspect is that physical maintenance measures alone cannot alter the social structure of a neighbourhood and the social problems that exist due to this structure. For long-lasting solution, social and physically oriented parties have to learn to work closely together.

The difficulties met in this phase are also closely related to the problems mentioned in the previous phases. In this phases it also important to obtain insight into the pros and cons of the environmentally friendly measures compared to the normal measures, especially for all the different parties involved. Knowing the different networks in a neighbourhood and the common and opposite aims of the participants of these networks can be very helpful to build bridges between the different parties involved (see also Teisman, 1992). Participation and the exchange of all kind of information are therefore aspects of great concern.

14.5   Solutions

Restructuring the post-war neighbourhoods needs to focus on solving liveability problems to today and, at the same time, on creating a new perspective for the future. During each phase of the planning process these two approaches need to be integrated and the difficulties in realising this need to be countered.

With respect to the problems occurring in the initiative phase we can point at the responsibility of the national government. The national government needs to give a sign to local municipalities and other actors involved in neighbourhood renewal processes that environmental aspect deserve more attention in investment than it receives nowadays. This could be done in many different ways. First, by stimulating the implementation of sustainable measures which have proven to be successful in other neighbourhood renewal projects by financial support. Second, by giving financial as well as expert support to neighbourhood renewal projects were the integration of sustainable measures is one of the leading strategies. Third, to set up a national library with information about implemented sustainable measures where one could find answers to questions such as: What are the effects of these measures on the environment, what are the pros and cons of these measures for the necessary functions for the neighbourhood? What are the costs and who is paying them? Who will receive the advantages and who will receive the disadvantages? These measures could be subdivided in different regions according to their geological and economic status. For example, for many water-based measures it makes a difference if the neighbourhood is build on sand or clay, and it makes a lot of difference if the neighbourhood is situated in an area were the housing market is booming or not. Figure 14.4 presents a very global method which offers support in finding the right ways to integrate sustainable measures in a neighbourhood renewal project.

Another recommendation concerning the initiative phase refers to the responsibilities at a local level. It is important to organise a stage for the actors; a platform on which they can influence each other. A project organisation can produce that platform and give the players involved the necessary authority in creating a plan supported by all the concerned parties. It is essential to realise that all the players have different goals and different roles to play. In a project group the dwellers and the experts should try to reach as much consensus as possible with respect to the existing and expected problems.

The plan also needs to be based on a vision for the future: a scenario that provides a context. The future of a neighbourhood depends on a large number of parameters, most of which are difficult to predict and cannot be influenced on the scale of the neighbourhood. A renewal plan, and therefore an ecological renewal plan as well, needs to fit in a scenario that is built out of a logical combination of exogenous developments. Those developments are sometimes directed by local policies and sometimes results from (inter)national economic and social developments. According to the preferred scenario or scenarios a programme that consists of a package of interrelated activities, measures, and projects needs to be designed.

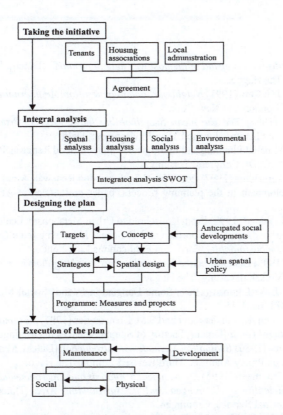

Figure 14.4   A planning method which offers the possibilities to realise sustainable measures in a neighbourhood renewal process

In Vinkhuizen, which was not a very large project, it was too difficult to work out all the of the described measures and projects at the same time and by the same persons. At this moment the announced measures and projects are divided between three project teams: one for social maintenance, one for physical maintenance, and one for the development of physical projects. A supervising team is responsible for the interaction between these teams and the end results.

The presented method will only work if all the actors are aware of the need of creating a new sustainable perspective for their post war neighbourhood and at the same time solving the liveability problems of today. Although there are probably no standard solutions, the exchange of experiences between ecological neighbourhood renewal projects is very important. Increasing knowledge about renewal processes and ecologically oriented measures enlarges the chance that sustainable measures will be realised in neighbourhood renewal projects.

Note

1   Dr. A. Bus is working at Saxion Polytechnic in Deventer, The Netherlands.

References

Agricola, E. (1995) *Who is afraid of the post war neighbourhood?* [Dutch], Foundation after the war, The Hague.
Alterman, R., and G. Cars (1991) *Neighbourhood Regeneration: An international evaluation*, Mansell, London and New York.
Bekker, F.H.W. (1994) *The European Sustainability Index Project: Project report*, The International Institute for the Urban Environment, Delft.
Bus, A.G. (1996) Spatial Planning and Ecological Neighbourhood Renewal: Paper presented at ACSP\AESOP Congress, Toronto.
Bus, A.G., and G.H. Heins (1997) Post War Neighbourhood Renewal: A mix of maintenance and development in the planning process, paper presented at AESOP Congress, Nijmegen.
Bus, A.G., and H. Voogd (1997) From Household to Urban Metabolism: Ecological renewal of neighbourhoods, in K.J. Noorman and T. Schoot Uiterkamp (eds) *Green Household Metabolism Sustainable?* Earthscan Publications Ltd., London.
Drontman, I.M. (1996) *Demolition and Renewal in Post War Neighbourhoods* [Dutch], Nirov-netwerk Volkshuisvesting, The Hague.
Etzioni, A., and E.W. Lehman (1967) Some Dangers in 'Valid' Social Measurement, *The Annals 373*, pp. 1-15.
Ketelaar, R., S. Kroon, K. van der Kuur and S.M. Mensonides (1997) *Sustainable Renewal (no or a) General Good* [Dutch], Faculty of Spatial Sciences in Groningen, Groningen.
Ministry of Finance (1986) *Evaluation Methods: An introduction* [Dutch], Ministry of Finance, Subdivision Policy Analysis, Third Revised Edition, The Hague.
Municipality of Groningen (1995) *Environmental-Analysis and Environmental Appreciation of the Municipality of Groningen and its Surrounding Area* [Dutch], Subdivision Environmental Service, Groningen.
Priemus, H., and G. Metselaar (1992) *Urban renewal policy in a European perspective, an international comparative analyses*, Delft University Press, Delft.
Sawicki, D.S. (1996), Neighbourhood Indicators: A review of the literature and an assessment of conceptual and methodological issues, *American Planning Association Journal*, Vol. 62, No. 2.
Teisman, G. (1992) *Complex Decision-Making: A pluricentric approach on decision-making about spatial investments* [Dutch], VUGA, The Hague.
Tjallingii, S.P. (1995) *Ecopolis: Strategies for ecologically sound urban developments*, Backhuys Publishers, Leiden.
VROM (1992a) *Trend Report Housing 1992* [Dutch], No. 43, Ministry of Housing, Planning and Environment, The Hague.
VROM (1992b) *Policy for Urban Renewal in the Future, 'the Second Half'* [Dutch], Ministry of Housing, Planning and Environment, The Hague.
VROM (1993), *Housing in Numbers 1992* [Dutch], Ministry of Housing, Planning and the Environment, The Hague.
VROM (1996), *Housing in Numbers 1995* [Dutch], Ministry of Housing, Planning and the Environment, The Hague.

Chapter 15

CiBoGa Site Groningen: A Breakthrough in Environmental Quality in the Densely-populated City

M. de Maaré[1] and E. Zinger[2]

15.1 Introduction

The Groningen Structure Plan *Stad van Straks* (The City to Come) reveals the great importance of the CiBoGa site. The local authority is aiming for a sustainable urban development and states that it wishes to make optimum use of the space available within the city boundaries. A high level of urbanism and a great variety of functions and urban cultures within these limits increase the practical value of the existing city. Building within the city boundaries also helps to limit automobility, particularly where this involves travel to and from work. The city therefore aims to construct 40 per cent of the housing target within the existing urban area.

The city does however lack large city-centre sites which are immediately available. Attention is therefore being paid to sites whose development has been prevented by circumstances for years. For example the CiBoGa site: the *Circusterrein, Bodenterrein and Gasfabrieksterrein* (circus, distribution and gasworks site), with a favourable location between the city centre and the earliest expansion areas and a favourable location as regards access routes and public transport and a relatively favourable environmental quality. But this sit has been neglected for years, with the remains of former industrial operations, some storage and parking. There has been a negative effect on the wider neighbourhood.

The site offers space for some 1000 homes and a wide variety of urban amenities. The development of this site has been in deadlock for many years, mainly because there is large-scale soil pollution under the site. For some time the local authority and province have been looking for a way to finance soil remediation. State funds for such work are however becoming increasingly scarce. The municipal policy is aimed at finding co-financiers for the remediation work (where possible the party who caused the pollution as well) and at reducing the remediation costs by harmonising developments below and above the ground. By working hand in hand with the national government and other partners, Groningen local authority feels it can break through the deadlock in development and achieve the added value that all the parties are aiming for: a sustainable, liveable and ecological urban area with a high physical planning quality.

15.2   Plan Area: A First Impression

The plan area is shown in Figure 15.1. The area to be built on is the CiBoGa site. The study area for the City & Environment project (see section 15.3) also includes the adjacent roads, as shown on the map. The CiBoGa site is located between Groningen City Centre and residential areas dating from the period of initial expansion. Boterdiep on the western side of the plan area forms part of an approach road from the north to Groningen city centre. To the south Boterdiep joins the so-called *Diepenring* around the city centre. *Nieuwe Ebbingestraat* is the major shopping street in the northern part of the city centre. The top of *Korreweg* has the function of serving the immediate neighbourhood. The site to be built on in the plan area measures just over 10 ha. The plan area occupies part of the former fortified walls demolished in 1874. While other parts of these walls were given a clear public function (e.g. the nearby *Noorderplantsoen* public gardens and the University Hospital) this site remained vacant for a long time. The most northerly part of the plan area was for years used as a Circus ground and retained this name.

On the largest part, the former gasworks site, the second Dutch gasworks was built in 1853. In the 1960s, with the introduction of natural gas, this site became redundant. The energy company did however remain on the site until 1995, surrounded by deserted industrial buildings. Apart from a number of monumental buildings, the majority of these have now been demolished. The distribution site (*Bodenterrein*), derives its name from the few hundred distributors who occupied the site between 1940 and 1970. The three parts of the circus, boden and gasworks sites are together called the CiBoGa site for short.

Round the edges of the site are various types of shopping streets. The *Kop Korreweg* shopping centre is deteriorating as a result of the failure to take action and the loss of support. Boterdiep has also suffered a similar fate. Nieuwe Ebbingestraat is struggling with a decline in customers. At present the CiBoGa site is itself largely vacant. A large part of it is used for parking, both for the university and the city centre. The long period over which the site has been in disuse has left its mark. The view of the backs of Boterdiep and Korreweg, combined with the partial lack of surfacing give the area a desolate appearance. Those living in the neighbourhood and passers-by feel the area is socially unsafe. This extensive use makes the area a forgotten island between the city centre and early suburbs, while the plan area should in fact, because of its location in the city, actually form a link.

In the meantime a few historic buildings on the Gasworks site have been taken into temporary use. For example, an exhibition hall has been installed in the old porter's lodge, showing the plans for the CiBoGa site. The site developer occupies a large villa where meetings and workshops are held. Another building has residential accommodation and houses the NNT (*Noord Nederlands Toneel* – North Netherlands Theatre).

15.3 Policy Framework: CiBoGa, Spearhead of Groningen Policy

The plans for the CiBoGa bring together many policy objectives. The local authority framework is described in the following.

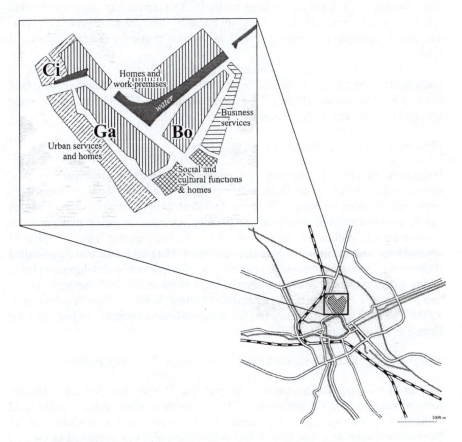

Figure 15.1 Situation of the plan area

'The City to Come' Structure Plan

In the 1995 'The City to Come' structure plan a careful balance was found between building in and around the city. Forty per cent of the homes to be built by 2010 will be built in the existing city. This fits in very well with the rising appreciation for living in the city as a result of changes in population composition and lifestyles.

Traffic and Transport Policy in Groningen

Groningen has a progressive traffic and transport policy. At present this policy is being updated in the project *Het kan verke(e)ren* (the title of which is a play on Dutch poet Bredero's motto 'It may change' and the Dutch word *verkeer* meaning traffic). The aim is to limit the growth of automobility to a maximum of 30 per cent in 2010. The aim is to have an excellent public transport system, plenty of space for cyclists and a directive parking policy. On the edges of the city, where the public transport routes cross the ring roads, there will be large car parks. To keep the city centre car-free but still accessible by car, parking garages will be built where the access roads cross the so-called Diepenring (see Figure 15.1). Car ownership and car usage in the city is relatively low.

Around 50 per cent of journeys are made by bicycle and 56 per cent of households in the city have a car. In the new ecological neighbourhood of Waterland (part of the Drielanden district) the parking standard is 0.7. The normal parking standard used for new building areas at present is 1.25.

Groningen Local Environmental Policy

Integration of physical planning and environment plays a great part in local environmental policy. The local authority has lately published a new (draft) environmental plan, *In Natura* (In kind). This draft plan considers at length how to handle environmental standards in the densely-populated city (see paragraph 4.8 compensation). Within the framework of local environmental policy a plan of approach for sustainable building is also employed. The plan focuses on the so-called 'Dubomatrix' which at all urban development levels (from urban development plan to letting and occupation) gives a number of fixed requirements and also a number of options (Local authority plan for sustainable building; What else!?, with dubomatrix, 1996). The local authority policy test for sustainable development can test plans for sustainability at an early stage.

National Policy Framework: the City & Environment Three Step Process

As part of the 'City & Environment' project, the Minister for Housing, Regional Development and the Environment (VROM) intends to offer around twenty local authorities the opportunity to experiment for five years with the so-called City & Environment approach. The City & Environment project was published in the 2nd national environmental policy plan and aims to find ways of dealing with the paradox of the densely-populated city.

Building in and around the city on the one hand offers unmistakable benefits from the point of view of environment and space, such as energy saving, efficient use of space and strengthening support for amenities such as public transport. On the other hand, this is sometimes made more difficult by environmental objectives for noise, odour, dust, soil and safety. The so-called City & Environment approach therefore endeavours to increase the opportunities for an integrated balance aimed at achieving an optimum quality of urban life. This is done in a number of steps. The first step: integrating environmental interests into physical planning at an early stage, must always be taken. Often the second step must also be taken: fitting and measuring within the scope offered by the present (environmental) legislation and regulations. Sometimes a third step is unavoidable: under strict conditions it is possible to deviate from the present legislation and regulations. This deviation may be both as regards content (e.g. standards) and procedure. The local authority has put forward CiBoGa as one of the experiments and is accepted as such by VROM. In this chapter, planning and the approach of the CiBoGa site are described in particular from the point of view of the VROM 'City & Environment' project.

15.4 A Sustainable, Low Traffic Density Area with Physical Planning Quality

The chance of again building a completely new area on the edge of the city centre requires more than an ordinary approach. In this new area, physical planning and environment are expressly inter-linked. One of the objectives is to achieve the optimum environmental quality. The area will be developed entirely on the basis of low traffic density. The starting points for a low traffic density area are particularly good: within walking and cycling distance there are all sorts of urban amenities and 30,000 high-grade jobs (University hospital, University, local and provincial authority, the city centre etc.). The location with respect to public transport, including bus and train, is also excellent. However, the creation of an optimum environmental quality in the plan area must not mean that environmental problems are shifted to the surrounding neighbourhood. For this reason the study area has been extended to the adjacent roads.

The CiBoGa site is extremely suitable for a densely-populated development and the program of requirements is also ambitious (see the draft plan of use in Figure 15.1):

- 1000 homes combined with work premises and small-scale social and cultural functions;
- over 1300 parking spaces, only 500 of which are for residents;
- 8000 m² large-scale retailers, including a supermarket on the Circus site and a development still to be worked out by the Boterdiep car park;
- 20,000 m² of offices/business premises for business services (university hospital).

Urban Development Quality

It may be anticipated that building on the CiBoGa site will have a positive impact on the neighbourhood. The support for the amenities present will be considerably increased, as will the safety and quality of life in the plan area and its neighbourhood. The urban development project reflects the history of the site. The contours of the former city wall contribute to the development of the public space, monuments and scenic buildings are integrated into the draft plan and the 'energy' aspect comes up in new construction on the gasworks site. The public character of the demolished fortress walls elsewhere in the city recurs in the draft urban development plan in the form of a number of high-density buildings (blocks) with a lot of open space between them. In building new homes the local authority is aiming both at existing and new city-dwellers, people from the immediate neighbourhood (Korreweg area) and those living outside who wish to return to the city. In this plan area housing types can be built which cannot be built elsewhere, if only in view of the high densities, of on average 70 homes per ha. The objective of reducing (auto)mobility also goes hand in hand with design requirements: some of the homes will have space for working at home (studios and workshops, offices). A number of experimental homes will be built with 'high-tech' facilities.

Traffic Quality

Boterdiep is one of the sites along the Diepenring designated in the local traffic policy for a parking garage. The access route from *Bedumerweg* is already a bottleneck. *Rodeweg* cannot now be used as a through route. With a view to accessibility this is necessary if only because of the autonomous traffic growth, which means fundamental reconstruction. The starting points for low traffic density development on the CiBoGa site are as follows.

A low traffic density development No through traffic routes in the area. This means that an existing traffic link, the *Bloemsingel*, will be lost . Within the plan area this means peace (no traffic noise and no exhaust fumes), space (for public green spaces, cyclists, pedestrians and playing children) and facilities for recreation immediately accessible from the home, such as walking and sunbathing.

The urban development design is aimed at letting as little traffic noise as possible penetrate from the adjacent streets to those functions sensitive to this, such as living and recreation outdoors. The housing designs can elaborate on this.

No parking at ground level For residential areas a parking standard of 0.5 is used. Parking costs are charged monthly, so that residents feel the burden of this each month.

The parking garages are concentrated in two places: at the top of Korreweg (supermarket and residents' parking) and behind Boterdiep on the gasworks site, geared to soil remediation (for residents and spaces for multi-purpose use: city centre visits, work at hospital and university, evening visitors). The parking garage on Boterdiep is combined with 6000 m^2 of large-scale retailers and is also called a 'source point' as it is expected to have a considerable impact on Nieuwe Ebbingestraat and the City Centre. Special attention is being paid to social safety in such a large parking garage. Also:

- Public transport routes will directly pass the area, with stops at suitable points.
- Special attention to cyclists: main routes and close-knit network.

- Additional measures to render car ownership unnecessary, such as a 'car pool system' in the area and additional service in the area of cycling, public transport, shopping etc.

Soil Quality

Within the CiBoGa plan area there are four cases which fall under the Soil Protection Act namely Cordes, Distribution site, former Gasworks site and the Circus site.

Cordes, filled-in Boterdiep The top soil (up to 1 m below ground level) of virtually the whole site is highly contaminated with lead. At occasional points in the soil, high concentrations of PAHs, chromium, copper and zinc are found. By the underground storage tank serious mineral oil contamination has been found in the soil and groundwater.

Distribution site, Bloemsingel Four relatively minor areas of contamination have been found on the site. By the former filling station the soil and groundwater are contaminated with oil and PAH; by the gas oil tank the soil is contaminated with mineral oil; by the carbolineum tank the soil is contaminated with PAH and in the centre of the site the soil is contaminated with lead and PAH.

Gasworks site, Bloemsingel The soil on this site is slightly to heavily contaminated with PAHs, cyanide and tar in the form of cinders and tar products. The soil contamination is within the site limits and varies in depth to a maximum of approximately 8 metres below ground level (by the former gasholders). The groundwater is heavily contaminated with cyanide, benzene and phenols.

Circus site, Korreweg The soil is slightly to heavily contaminated with PAH and mineral oil at four points. The groundwater is heavily contaminated with PAHs, mineral oil and aromatics. The complex position as regards soil contamination is a technical, but primarily a financial problem. In addition, the many formal procedures required for the soil remediation process are very time consuming. The starting point for soil remediation is a function-oriented approach, aiming for maximum harmonisation with the above-ground plan development. While drawing up the urban development plan, a soil remediation option has been worked out for the whole area, within the limits of the Soil Protection Act. A decisive factor in the plan is the contamination of the former gasworks site. At the most contaminated point a single-storey parking garage is being built so that the contamination is isolated in the horizontal depth by the floor of the car park. Sheetpiling around the parking garage provides a vertical barrier to the contamination. Finally, the parking garage has another important function as an 'intermediate layer' under the building, to eliminate health risks from evaporation as far as possible. In the remainder of the plan area, outside the barrier, all the contaminated soil and groundwater undergoes multi-functional remediation.

Further Environmental Quality

Contrary to other parts of the city of Groningen, the CiBoGa site itself has a low environmental load from surrounding activities. At an urban level there are no sources of pollution, apart from the serious soil contamination and the traffic noise.

Traffic noise represents one of the paradoxes, but also one of the challenges of the plan. The CiBoGa site will have a low traffic density, but in the immediate vicinity traffic flows will increase considerably. In the case of Rodeweg, the local authority is hitting the limits of the Noise Nuisance Act. The roads bordering the CiBoGa site at present have an intensity of around 10,000 traffic movements per day. Because of autonomous traffic growth, road restructuring, the economic prosperity of the district and the residents of the CiBoGa site, these intensities will increase by 5000 to 10,000. The biggest increase is anticipated on Rodeweg, from the present 500 to around 20,000 traffic movements per day. This is because of reconstruction, with an increase in noise which is considerably above the increase of 5 dB(A) permitted by the Act. This affects approximately 80 homes. It is not expected that the compensation required by the Act in the form of a reduction in homes suffering nuisance in the neighbourhood will be possible. On Boterdiep, around 110 homes will suffer an increase in noise from 63 to 65 dB(A).

At present no air quality limits (benzene, CO and NO_2) are exceeded in and around the CiBoGa site. In spite of the increase in traffic as a result of the planned parking garages, it is anticipated that the situation with regard to air quality will not get any worse. A number of design and technical measures are however necessary for this, for example a change in profile and underground entrance in the central reservation. The parking garages themselves have mechanical air extractors, with an outlet point at a suitable location on the site.

On Boterdiep there are a few businesses suffering nuisance. It is anticipated that the problems will be solved. Work is being carried out to obtain a permit for a youth centre, for which sound-proofing measures must be taken.

The new building on the Gasworks site will take into account the permit granted to this youth centre. (New businesses causing environmental nuisance are not permitted in the area.)

Sustainable Quality

Since this is the planning phase, so far sustainability aspects have been looked at particularly on an urban development level.

Energy It is assumed that all the homes will be built below the legal standard for energy-use applicable at the time of construction. Because of the density of the site, a combined heat and power solution seems to be the obvious option. The energy company (ESSENT) has investigated how to make use of the spare capacity of the nearby combined heat and power plant at the hospital. ESSENT is considering a micro heat and power plant, where one boiler is required per 5 to 10 homes. It may then be possible to lay a mini gas network to the homes for cooking. Supplementary heat storage from solar energy is an option.

Waste collection A form of waste collection will be implemented in the area geared to the low traffic density and the high quality of the public spaces aimed for. This means a method with as little disruption to public spaces as possible and a minimum of transport movements.

Other sustainability aspects, such as water consumption and the use of materials, will be further elaborated in the housing design stage. The local authority plan of approach for sustainable building forms a starting point here. In addition to these specific items, it is important to pay attention to flexibility. The building is not only for the present, but for the future as well. Consideration should be given both to being able to take advantage of future technological developments (infrastructure) and social developments (use of homes/businesses and public spaces).

Ecological Quality

The development of the CiBoGa site will form an important corridor in the urban ecological structure. The Noorderplantsoen public gardens are now in an isolated position within the urban ecological structure. The value of the Noorderplantsoen derives from the variation and age of the trees. Efforts are therefore being made to create a continuous ecological corridor between the gardens and the new urban district park in the form of rows of trees. This will provide better opportunities for the movement of birds and bats present in the gardens. In part, these trees are already present in the area. Use can also be made of the 'moving' of a number of trees from the Noorderplantsoen which has already taken place. In particular the section by Rodeweg is a fragile part of this corridor: trees must as far as possible be retained in the little space available. The public space in CiBoGa is designed and managed on an ecological basis. A favourable factor in the green space structure is that the percentage of paving on the CiBoGa site can be kept low. We are working on 16 per cent instead of the 22 per cent normally used. This means additional water infiltration is possible.

The present quality of the water in the Noorderplantsoen is moderate. There is only one underground link with the Oosterhamrik canal. The intention is to improve the existing water quality in the years to come. Work is already being done on this in the Noorderplantsoen, to increase the quantity of buffer space in the area and to make a water link with the Oosterhamrik canal. If necessary a separation will be created from the more polluted water of the Starkenborgh canal. These measures will create clean surface water in the plan area. On the northern side of the Oosterhamrik canal, a nature-friendly bank can be created, with marshy vegetation. On the sunless side of the canal (where the old city wall was located) there are opportunities for wall vegetation. As part of this corridor a new lake to be excavated on the Circus site will play a part in improving groundwater quality.

Via Compensation to Environmental Quality

The above shows that the environmental aspects can largely be solved within step two of the City & Environment project; that is, within the existing range of environmental standards. This will be difficult enough. As regards the environmental standards, it is anticipated that only Rodeweg will cause a formal problem with the Noise Nuisance Act.

If Groningen local authority can implement its plans for the CiBoGa site, ideas will have to be worked out regarding compensation. The CiBoGa project sails under the flag of one of the projects from *In Natura* (In kind, the local environmental policy plan), namely 'Dealing with Standards'. A 'compensation method' is being developed here on the basis of five pilot schemes. The 'Dealing with Standards' project came into being thanks to a working group whose members included representatives of the physical planning service, the environmental service and the emergency services, as well as the province and the regional environmental health inspectorate. The working group was at the same time working on putting together the City & Environment policy framework and took the three-step method as a starting point for looking closely at the situation in the city. One of the conclusions was that there may also be justification for considering compensation measures to improve the quality of life within step two. This may for example be the case if the objectives for an area are very much aimed at environmental quality, or if the residents in an area are, without being asked, faced with a higher than normal increase in environmental nuisance. The CiBoGa site involves both these factors.

Groningen local authority will involve the University of Groningen in the development of a compensation method. An initial draft of the compensation method is at present being considered. This will be studied further in the years to come and its practical feasibility tested. An example of an application of this approach to CiBoGa is compensating for the construction of parking garages (which involve additional traffic movements to and past the CiBoGa site) with a low traffic density development. So compensation – following the City & Environment policy framework – is in the first instance sought within the same sector, then within related sectors (e.g. soil, water, ecology and noise, energy, waste) and finally – only where no other ways are open – by social or physical planning measures. Of course compensation options will also be considered for the new situation of Rodeweg, and possibly the other adjacent roads.

Monitoring

In the course of the planning process a monitoring system will be worked out incorporating both performance-oriented and effect-oriented items. One can think of:

- noise – the actual increases in noise nuisance on for example Rodeweg and Boterdiep and the level of freedom from noise achieved within the CiBoGa site;
- the number of parking places created;
- energy – the number of homes built under the legal standard for energy-use;
- water – the level of improvement in water quality in the area, the construction of a water link between Noorderplantsoen gardens and Oosterhamrik canal;
- green spaces – the creation of a continuous green corridor; and

- waste – the form of waste collection. In addition one can also think of monitoring perception, both for residents from the neighbourhood and new residents.

15.5 The Process

The CiBoGa site has for years been a 'black hole' on the edge of the city centre. Even when the local authority carried out the first soil inventory, in 1980, the gasworks site was mentioned. In subsequent years initial inventory surveys of the soil were carried out. There has been uncertainty about the future of the former gasworks site, because of the question of the liability of the energy companies in The Netherlands. This also prevented adjacent sites being developed. In spite of this, the local authority had already developed plans for the area. At the end of the 1980s the plan was produced for a low traffic density area, and was greeted with enthusiasm by the State. Only after agreement was reached for the energy company to move out could the plans be pursued seriously. This was in 1993. From the outset it was clear that as well as a remediation plan an attractive urban development project was wanted. With the province, which is responsible for the soil remediation, a strategy was developed for talking to various ministries. A memorandum on the starting points was written, as a precursor to the urban development plan.

In 1995, the market parties signed an agreement with the local authority and together set up a fund from which the planning costs are paid. A firm of consulting engineers was called in for the soil remediation plan. In the course of time a complex project organisation was created, with various focal points. As this is such a special plan, careful consultation must be carried out and creative consideration given to many areas. The central figure for all the consultation bodies is the project manager who is responsible for harmonising the different policy fields. But to achieve actual harmonisation the various disciplines in the project are also frequently invited to take a look at one another's work. For this reason the design team also included soil experts. On a number of occasions there has been consultation with the State at different levels, particularly to find a solution for the soil problem.

A so-called open planning process is being used. In the design phase workshops were held at regular intervals, in which internal and external experts could discuss the plan, or parts of it, with one another. A communication plan was drawn up, in which consultation within the neighbourhood takes shape in the form of a residents' and a business panel. The local authority facilitates this by financing an independent chairman and secretary, and providing meeting premises. Information evenings are organised at regular intervals.

At the moment the finishing touches are being made to the urban development plan. The planning process is at the same time aimed at getting a commitment from the (state) partners. To speed this up, a financial strategy is followed which is not simply based on regular project contributions. Several financiers are sought, such as the market parties and the energy company. Opportunities are also being sought to remove the divisions that exist between sources of finance. Switching between short and long term investments would have to be possible. This involves creating win-win situations: all the parties must gain something. The City & Environment project may possibly be able to extend the opportunities for this.

There is also a lot to watch for with regard to the required procedures. The present basis for the plan is the Structure Plan, in which the function and site (CiBoGa) are established. Procedures involved concern for, among other things, preferential right, expropriation, soil contamination, soil remediation, noise nuisance, air quality, withdrawal of roads, monument permit and land use plan.

The intention is in the preliminary stage to organise the basis and quality via the open planning procedure and hence to speed up the formal stage. Benefits can also be derived from joining together and combining procedures. The local authority wants to prevent unnecessary time being required once there is a breakthrough in the decision-making process. The open planning process must offer scope for sufficient administrative commitment and social support.

15.6 Conclusion

In implementing the CiBoGa plan sustainable urban development will take specific shape. The area must be designed with a low traffic density and without through car traffic. By tailoring the work in the planning phase, the plan area will remain the environmental nuisance-free site it now is as regards traffic. An additional challenge is to ensure that the larger urban area is not saddled with additional environmental problems. Energy, ecology and waste collection are spearheads in the sustainable character of the area. By concentrating urban functions into very compact modules, the plan can combine a high density with a spacious layout. At the same time this is a chance for pronounced urban housing types with a particular architecture. Groningen hopes to soon resolve the financing of the plans and to make a start on execution.

Notes

1   M. de Maaré works as a policy officer at the Environmental Agency of the Municipality of Groningen, The Netherlands. She graduated as a social geographer and an environmental specialist. Her responsibility is to integrate environmental aspects in urban town planning.
2   E. Zinger is partner at the 'Nieuwe Gracht' consultancy in Utrecht, The Netherlands. He works on urban development and the quality of the environment.

References

Municipality of Groningen (1995) *De Stad van Straks, Groningen in 2005*, Municipality of Groningen, Groningen.
Municipality of Groningen (1996) *Wat anders!?*, Local authority plan for sustainable building, Municipality of Groningen, Groningen.

Municipality of Groningen (1997a) *De bereikbare stad leefbaar*, Traffic Plan for the City, Municipality of Groningen, Groningen.

Municipality of Groningen (1997b) *In Natura*, Environmental Plan for the City 1997-2000, Municipality of Groningen, Groningen.

VROM (Ministry for Housing, Regional Development and the Environment) (1995) *Waar vele willen zijn is ook een weg, policy frame City and Environment*, VROM, The Hague.

Chapter 16

Sustainable Downtown Urban Renewal: Redefining Yonge Street

J. Oosterveld[1]

16.1 Introduction

For several reasons the urban core is one of the most challenging areas for dealing with environmental problems. First, it does not embody the urgency of a contaminated industrial site nor does it command the emotional response that a residential neighbourhood may inspire. Due to its commercial orientation, the environmental health of the urban core is ultimately equated with its economic vitality. Thus, in an economically challenged downtown area, the natural environment does not hold priority in renewal efforts. Finally, due to the physical density of built form and impermeable surfaces, little more than the weather acts as a connection to the natural world. Many people including consumers, residents and business owners remain apathetic to their environmental impact within the urban core.

An environmental cleanup of the core requires more than simply planting a few trees. Rather, it requires physical renovations as well as a major change in social habits and business practices. This chapter investigates Yonge Street in Toronto which is characterised by vehicular congestion, along with commercial and pedestrian areas that are in economic and social decline. The corresponding urban renewal project, the Downtown Yonge Street Regeneration Programme, addresses these problems simply by cleaning the building facades and sidewalks as well as targeting popular chain stores to locate. This solution does not offer, in my opinion, a substantive base for economic recovery, nor does it recognise any type of commitment to environmental issues and community character.

Urban renewal requires a revolutionised use of space through both physical design and resulting usage in order to be sustainable over an extended period of time. The major hurdle for this type of project is the narrow perception of what 'renewal' should encompass by those who usually initiate these projects. A vibrant street is made up of a number of subtle elements that reach far beyond what is sold in its stores. Broadening the scope of renewal allows environmental activism to become a valued component of the project. On the other side, urban renewal will become a vehicle for environmental improvement by activating those who have the most influence over the street namely, the property owners, business people, residents and the municipality.

This chapter aims to provide an evolutionary approach to community-oriented urban renewal projects based on sustainable development which includes social, economic and environmental issues. The specific case study chosen for this chapter is the Downtown Yonge Street Regeneration Programme which targets the core area of Toronto along its main street. The Regeneration Programme is focused on re-establishing the street's pre-eminence within the city in light of its 200th anniversary. The recommendations of this chapter intend to set the project's focus into the future so that its 300th anniversary can prove prosperity and sustainability within the urban core.

This chapter will first highlight issues of economic growth and environmental sustainability in the downtown core in their present context. The focus will then review the present downtown Yonge Street revitalisation efforts which will be followed by recommendations that will consider long-term economic and environmental sustainability in the context of community participation.

16.2   Downtown Urban Renewal Projects

Vitality of the urban core, especially along its main street, is essential to the character and dynamics of the city. It not only leaves a strong impression on visitors, but also contributes to the health of the entire urban area. The importance of a downtown core is undisputed among authors as the backbone of the city's identity (Robertson, 1995; Barnes, 1986). The city's identity, however, is based on social, economic and cultural criteria while ignoring its environmental health. 'Despite three decades of continuous redevelopment policies and projects', Robertson cites that the image of the urban core still remains 'inconvenient, obsolete and dangerous' (Robertson, 1995, p. 429). To alter these negative perceptions, other authors suggest that a sense of pride and responsibility in the city's identity should be reflected in renewal projects (Gruen, 1986). Gruen places an emphasis on a synergistic approach which works at both a municipal level and a community level through the participation of local business owners and residents.

One challenge for urban renewal involves maintaining a unique identity while attempting to enhance economic vitality. Present trends in project strategies tend to follow a perceived formula for success that leads to similar features in cities around the world. Robertson lists seven widely used strategies throughout the past three decades: pedestrianisation; office developments; special activity generators (for example, convention centres) and transportation enhancement (Robertson, 1995). He notes: 'now that redevelopment strategies have been applied similarly in cities across the country, uniformity of function and appearance blurs the distinctiveness of each downtown' (Robertson, 1995, p. 436). An example of this scenario is the recent development of waterfront casinos as found in Melbourne, Australia; Adelaide, Australia; Niagara Falls, Canada; Windsor, Canada; and many other North American cities. Main street revitalisation has more recently been the new target for medium-sized and big box international corporations including Nike, Sports Authority, the Gap, Planet Hollywood and Virgin Records in cities such as Chicago and New York as well as Toronto's Yonge Street (Sewell, 1996). Urban areas seeking economic renewal welcome their clean, popular corporate image in order to increase the standard of quality among businesses in the area. Main streets are thus faced with an identity crisis, striving to maintain their definitive characteristics while attempting to shift out of economic depression.

A successful downtown has been described as 'full of surprises – unusual shops, theatres, galleries and businesses' (Gorrie, 1991, p. 70). Ghiardelli Square in San Francisco aimed to address the issue of distinction among urban experiences by recognising the need to renew the physical appearance as well as planning marketing strategies that target both businesses and customers (Barnes, 1986). Barnes attributes part of the long-term success of Ghiardelli Square to the special events which offer flexibility and can respond to market fluctuations without requiring a permanent change in infrastructure. The management also made efforts to attract stores unique in variety and style with 33 of the 40 new stores as non-chains (Barnes, 1986, p. 9). This development also takes pride in its industrial heritage. The conversion of these buildings into their present commercial use has created a unique space which is complimentary to the non-chain shops while providing an experience that only San Francisco can offer.

In summary, a number of prevailing themes arise throughout the research on the economics of urban revitalisation that should be incorporated into the proposed vision of urban renewal. First, maintaining the unique qualities of the streetscape must be considered a priority and the 'suburbanisation of downtown' avoided (Robertson, 1995, p. 436). Multifunctional spaces need to be considered in the design of both public areas and private businesses. Public spaces allow for a variety of promotional and community activities to attract people to the area. These activities will vary seasonally and according to current trends. The design of commercial space may require, as in the case of Yonge Street, the amalgamation of properties in order to accommodate the larger retail operations. By requiring the quality of flexibility into their store design, the community is able to ensure continuity in response to changes in retail occupancy. Foreseeing and designing for the potential closure of a larger chain store can help the street remain less dependent on these sanitising invaders.

16.3   Urban Environmental Issues

Sustainable development is defined by the Brundtland Commission as the means to 'meet the needs of the present without compromising the ability of future generations to meet their own needs' (as quoted in Nickerson, 1993, p. 15). Globally, Agenda 21 was established at the 1992 United Nations' Earth Summit to ensure that sustainable development policy is integrated into local Official Plans by the end of 1996. Environmentalism now must move from the pages of official documents and into action.

Toronto's Official Plan (Approved 1993) has incorporated an environmental mandate as part of its *Livable Metropolis* theme which aims to balance a healthy environment with economic vitality and social well-being (Metropolitan Toronto, 1992). The development of the Plan emphasised the importance of sustainable development within the urban setting as 'one of the City's most pressing public issues':

> In the 1990s, the concept of balanced growth must not only acknowledge and
> have regard for the environment and environmental concerns, but it must fully
> integrate environmental consideration into the land use planning and
> development process. (Metropolitan Toronto, 1991, p. 9)

Although the Official Plan emphasises the importance of social, economic and
environmental issues, it does not recognise their symbiotic relationship in the form of
integrative policies.

Environmental policies, generally, equate ecology to soil. For example,
Toronto's Official Plan primarily discusses environmentalism in the context of the
protection and rehabilitation of lands, especially natural corridors and old industrial
sites. Creating sustainability along Yonge Street becomes challenging due to its soil-
less nature. Seas of pavement abutting dense building structures leave little visual
association between humans and nature. In analysis of a cross-section of the street one
begins to recognise the intricate range of environmental impacts that exist within the
core (Table 16.1).

Table 16.1   Environmental problems of the urban core

Realms of responsibility	Examples of environmental issues
Vehicular street	• CO_2 vehicular emissions • Lack of bicycle lanes • Runoff of salt, oil and gasoline into the water supply
Pedestrian realm	• Lack of recycling facilities • Harsh climate • Lack of bicycle lock-up facilities • Lack of trees and permeable surfaces
Private buildings	• Energy inefficiencies • Off-gassing from building products • Excessive throughputs to infrastructure (garbage, water) • Sale of high environmental impact products

A study of urban municipalities in England and Wales endeavoured to
understand why sustainable development has not been incorporated into mainstream
local action (Gibbs et al., 1996). Gibbs, Longhurst and Braithwaite concluded that the
environment and the economy are perceived as a simple dichotomy. In this case, an
either/or situation exists. This is because the economy, whose relationship with
employment seems more tangible in comparison to abstract environmental concepts,
tends to marginalise nature as simply traditional park space (Gibbs et al., 1996). This

pigeon-holed perspective of the environment ignores that the economy functions within the limitations of nature.

Respondents in the England and Wales survey felt that lack of finance and other priorities (mainly, employment) were the main obstacles that placed the environment as a lower priority. One local authority respondent successfully summarised the overall problem with this statement:

> Confidentially [...] there is so little appreciation and understanding of environmental issues and what their incorporation requires or could mean, that all I can say is that we have a very long way to go. It only happens (or rather it is mentioned) when doing so is a requirement for getting money e.g. European or Department of Environment grants or budgets, and making 'caring for the environment' statements in corporate plans. (Gibbs et al., 1996, p. 327)

The general apathy towards the environment is further justified by the long-term pay-back time related to its integration. In an interview with Danny Bellissimo, a planner for the City of Toronto's Economic Development Division, he noted that the greatest obstacle in for environmental proposals is the political viability or short-term gains required to prove a change is taking place during a politician's term in office (Bellissimo, 1996). The Official Plan, he explains, becomes 'lip service' while short-term gains, especially, employment and economic initiatives take political precedence. He also expressed the reality of 'passing the buck upwards' to provincial and federal levels of government when attempting to resolve who will pay for these efforts. The challenge then becomes designing a renewal strategy that incorporates short-term political or economic gains while working towards long-term environmental improvements.

16.4   Yonge Street, Toronto

Yonge Street has been the focus of revitalisation efforts since the early 1970s. The Chief Planner to the Metropolitan Toronto Commissioner of Roads and Traffic in 1970 explains that Yonge Street's character has been established in the public consciousness as Toronto's main street. The street maintains a special significance due to 'the street's centrality, length, accessibility, historical importance, visual and physical crowdedness, and as the principle shopping and entertainment street' (Toronto Planning Board, 1970, p. 5). A recent project focused on Yonge Street's 200th anniversary in 1996. The Yonge Street Business and Residents Association Inc. (YSBRA), in co-operation with the City of Toronto Urban Development Services, established the Downtown Yonge Street Regeneration Programme as a three year project to enhance the 'quality of life' of Yonge between Gerrard and Queen Street (as located in Figure 16.1). The main goal of the project is to increase the number of visitors and residents as well as the proportion and range of businesses who locate on Yonge Street through: the improvement of safety, appearance and cleanliness of buildings and public space; a promotion campaign to attract customers; a marketing campaign to attract quality retail and entertainment facilities; and, finally, a strategically focused redevelopment program which aims to improve the quality of the street environment (YSBRA, 1996, p. 2). Through the set of principles and guidelines entitled Downtown Yonge Street Improvement Plan, Council approved the plan aimed to make 'improvements to building facades and the streetscape to assist the economic revitalisation' (City of Toronto Planning and Development, 1995, p. 6).

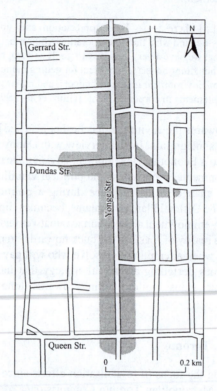

Figure 16.1 Focus area of Yonge Street
Source: Planning and Development Department, 1995

   The first stage of this project designated the downtown area of Yonge Street as a Community Improvement Area. One of the target projects under this designation involves the City of Toronto's Commercial Facade Improvement Loan Program which provides interest-free loans for the first twelve months, with a below market rate of interest to be charged for the rest of the loan term. The Downtown Yonge Street Commercial Facade Improvement Grant and Loan Program encourages business owners to take responsibility for Yonge Street's face-lift while preserving historically significant sites. The Eaton Centre has also budgeted moneys for public realm improvements on its private property, which include tree planting along Dundas Street West, an outdoor café with human-scale lighting, as well as hanging flower baskets along Yonge Street (City of Toronto Planning and Development, 1995, p. 14).

Marketing campaign efforts have targeted large chains, many of which are considering the move to Yonge Street including the Sports Authority, the Gap, Universal Studios, Nike, Crate and Barrel, and Tower Records (Sewell, 1996). City Councillor Kyle Rae is promoting these large International businesses to locate on Yonge Street (Barber, 1996). Ronald Soskolne of the YSBRA, however, recognises the need to reduce the sanitising effects of these businesses (Barber, 1996). The project's challenge will involve developing a context for these businesses to locate in order to maintain the unique qualities of Yonge Street.

16.5   The Sustainable Street

In order to consolidate a strategy for urban renewal, Table 16.2 summarises a list of goals as derived from the literature as well as through interviews with parties involved with Yonge Street. These goals provide the driving forces for the renewal model. This model, presented in Table 16.3, is based on three main principles. The first principle requires a broader group to be included in renewal efforts from the earliest stages of the project. The aim of this principle is to begin to diversify the issues that the project is able to address by adding a broader range of expertise, creativity, skill set, and human energy. This diverse group can then establish a theme for the project that is specific to the local community that attempts to encompass the goals of the participating parties. This theme is meant to give a context for new marketing strategies, physical renovations, and community programming as a means of inviting the community back to the street. Maintaining the theme becomes part of the final principal of this urban renewal concept, namely, longevity. By considering renewal efforts as the first stage in an evolutionary process, the street can potentially work to avoid a roller-coaster ride of renewal efforts and economic slumps.

Table 16.2   Goals for downtown urban renewal

1   *Long term* economic, social and environmental sustainability
2   Establishment of a *unique street experience*
3   Development of the *main street's pre-eminence* within the community
4   Increase *pedestrian activity*
5   Attract *new businesses* to the street
6   Create a sense of *pride and responsibility* for the street within the community
7   *Educate* the public on the environmental impact of a commercial area
8   Provide an active means of carrying out the intent of the *Official Plan*

Forever Yonge

This section lends example to Yonge Street in the context of the model presented in Table 16.3. The YSBRA was established at a grassroots level to bring about economic renewal for Downtown Yonge Street. Some of their limitations included financing, time constraints, and human resources (Sniderman, 1996). By inviting a broader range of community participation including representatives from Toronto Green Communities Initiatives, local student groups from Ryerson, University of Toronto, and social welfare organisations, the group is capable of addressing a wider range of issues that affect the street.

In order to understand the local context, a street audit should be completed in the early stages of planning in order to evaluate environmental issues as well as personal safety, aesthetics, and other concerns. Targeting realms of responsibility including the vehicular street, the pedestrian realm and private businesses will make concerns more manageable for each participating group. The renewal group can then review these concerns to establish its mission statement, theme and goals to ensure that the issues can be addressed effectively.

The Yonge Street Business and Residents Association's present mission statement provides the fairly obvious goal of re-establishing the street's pre-eminence in the community. It is, however, overly limited in scope. Environmentalism or sustainable development provides a progressive tool for change and long-term stability. Yonge Street's current renewal project slogan *200 Years Yonge* can become a commitment to the future through the theme, *Forever Yonge*. The mission statement, as shown below, has been amended to include sustainable development as the means to achieve their goal:

> The YSBRA is a proactive, representative mix of small and large Yonge Street Businesses and residents and other community members working for the long-term reestablishment of downtown Yonge Street as Canada's pre-eminent retail business, cultural, residential and social district through their commitment to sustainable development practices. (YSBRA Newsletter, 1996 with amendments)

Table 16.3   Working toward sustainable downtown renewal: General guidelines for grassroots initiatives

Initial stages of the project:

1	Form a broader group	(human energy, creativity, and expertise)
2	Street audit	(localise the issues)
3	Mission statement	(establish a theme for renewal)
4	Vision of the street	(specific means to work toward the goals)

Evolutionary process:
Immediate & Long-term Changes Through:

5	Vehicular street	(municipal renovations)
6	Pedestrian realm	(municipal renovations & community programs)
7	Private business	(municipal incentives & business initiatives)

Table 16.4   The sustainable street as created by a combined effort

Action Taken by:	Main Motivation for Participation:	Potential Action:	Results:
Community Groups	Educate the public about the group's agenda (e.g. environmentalism). Raise its public profile. Create mainstream activism.	Plan activities which promote the street's theme and group's agenda including: Earth Day festival. Design competition. Street tree planting. Bike tour.	The reintroduction of the community to the urban core. Higher customer volume for local business. Public awareness of renewal theme and its global reasoning.
Businesses	Economic growth. Encourage new businesses into the area in order to attract more customers to the area.	Aesthetic improvements complimented by efficient business practices. Promotion of project theme through: Business tours. Sales and discounts	A collective theme for business owners to advertise. Media attention. Higher volume of customer activity. New businesses will take advantage of the street's theme.
Municipality	Show the City's commitment to the Official Plan. Political gains.	Add environmental constraints to funding incentives and other policies and guidelines. Make infrastructure more sustainable.	Proof of political commitment. Provide assurance for businesses to take the risk. Potential economic savings on infrastructure.

In order to avoid the mistake of sustainable development becoming a generic form of revitalisation, as in the case of other trends, the renewal group must work to establish a theme in the context of the community's prominent character. Yonge Street as a leader in main street sustainable development fits its position of pre-eminence among Canadian cities. Its context is further solidified by connecting its place in history with its place in future development of the city.

Setting goals and design criteria will further bring the abstract mission statement into practice. The key goals of the sustainable Yonge Street are highlighted in Table 16.2. Each group participating in the project will have their own self-promoting agendas. Common goals will provide a basis to work from so that the activities of one group will compliment those of another as shown in Table 16.4.

On Yonge Street, improvements to the vehicular street would be addressed by the municipality, the pedestrian realm by community groups and the municipality, and private businesses would work in co-operation with the municipality. First, changes to the vehicular streetscape would create, in my opinion, one of the most dramatic changes to the street's character. Currently, Yonge acts as a thoroughfare with four congested lanes of traffic. Not only would a reduction in traffic to two lanes change the volume of vehicles that use the street, but it would also create a healthier, more comfortable pedestrian environment. The additional space could then provide an equal priority to lower impact forms of transportation compared to the coveted automobile.

Groups in Toronto that may be interested in working on a project such as this could include Healthy City Toronto, East Toronto Green Community or the Toronto Atmospheric Fund. Healthy City Toronto has done research into the use of bicycles in the central area. Their research concluded that there has been a 270 per cent increase in bicycle use coinciding with a 25 per cent decrease in motor vehicle use in the central area since 1976 (Healthy City Toronto, 1991, p. 56). This research can add justification to such a controversial street renovation. As part of the evolutionary process, changes must occur in succession in order to avoid substantial up-front costs. The first stage in the process is to change the outside lanes into priority lanes for public transportation and bicycles. The long term changes will then take place as part of the next road works project. This phase involves narrowing the road to two vehicle lanes and two bicycle lanes so as to provide additional space for pedestrians.

Changes to the physical street would further affect the pedestrian environment. Municipal renovations to this area should consider the life-time impact of the product installed. Street lighting, for example, can act as a showcase for new and unique low-impact systems which may feature energy efficient bulbs, recycled material, and overall creative designs that maximise lighting output with minimal resource inputs. Such considerations would exemplify the City's commitment to sustainable development as proclaimed in the Official Plan.

Community programming will further aid in generating pedestrian activity on the street. Members of the community who do not make a habit of visiting Yonge Street need an incentive to return. This task in itself would be quite difficult for the business owners without the help of the broader community. A vibrant main street provides more than just economic benefits to local business. Community groups can use the renewal project as a means of their own self-promotion while establishing unique programming that enhances to the street's character. Sustainable Yonge Street can act as an educational tool and a place to celebrate the city. Programming for Yonge Street is able to feature events around Earth Day, or international conferences such as the Humane Village Conference. In addition, events that help to 'rebuild' the street will also create a community sense of responsibility and control over the development of their main street. Events for Yonge Street could include a street tree planting, a demonstration roof-top garden or a design competition. The design competition would actively involve students at University of Toronto, Ontario College of Art and Design and Ryerson Polytechnic Institute with an actual physical project. The design focus would reflect the theme of sustainability while adding unique features to the street for example, a competition to create street furniture out of recycled materials.

The third realm of responsibility, the private business sector, involves both private initiates as well as municipal incentives. Since not all renewal efforts will prioritise environmental issues, municipal incentives need to be supported by guidelines to control how public funds will be spent. The Community Improvement Area, for example, defines its mandate to include the improvement of 'amenity, appearance and environmental quality' (Metropolitan Toronto, 1993, Section 15.2.a). A more concise definition shifts 'environmental quality' from an aesthetic term to one that addresses ecological impacts. Further, the Commercial Facade Improvement Grant and Loan Program also needs an environmental mandate or design criteria to ensure that renovations use energy efficient windows and lighting, as well as recycled, reused, non-toxic materials.

Environment-oriented community groups can help to educate businesses on low-impact building products for their renovations. They may also have other suggestions on reducing the impact of their business practices for example, requesting that manufacturers reduce the amount of packing materials; using less paper during sales transactions; and selling low-impact products. Business owners must recognise that renewal requires dramatic action in order to achieve a substantial overall change in the streetscape. The businesses on Yonge Street can benefit from the sustainable street image through the activity generated from community programming and the resulting unpaid advertising of media attention. Business owners can become further connected to the sustainable image by designing their promotional activities around community events for example, a discount for those who rode their bicycles downtown, or store tours that feature positive environmental efforts on their part.

With an established street agenda, both small entrepreneurial and larger established businesses will be attracted to Yonge Street for its positive image. The street's reputation will act as a form of promotion and unpaid advertising for their own benefit. Small businesses can take advantage of the fragmented nature of the properties along Yonge while adding character and uniqueness to balance with larger chain stores. These small businesses can provide a mix of unique products that will attract a wide range of people and encourage economic diversity. Unique environmental businesses will further enhance the street's image for example, a café with a roof-top garden, or a clothing store with products produced from fabric ends. Larger, well known businesses also need to incorporate an environmental mandate in order to take advantage of the sustainable street image. The Gap, for example, has taken steps to improve their corporate campus in San Bruno, California by working with ecologically and economically innovative products including a native grass roof covering, and natural lighting systems (O'Connor, 1996, p. 86). These principles can and should be applied to their store locations. Sustainable Yonge Street would be a perfect location to feature their progressive behaviour.

16.6   Summary and Conclusions

In both industrialised and developing countries, urban pressures are growing exponentially. Issues including excessive consumption rates and urban sprawl need to awaken us to the pattern in which our cities are evolving. Those responsible for the growth pattern of cities must begin to look inward and design more efficient urban systems. We must learn to take responsibility for our environmental impact as individuals and as a community. This concept, however, still remains intangible, especially in the urban core whose physical structure seems far removed from any natural ecosystem. There is a limit to what environmental activists can accomplish unless they have the co-operation of those who control the land. Urban renewal projects offer an ideal means for environmental action in this area since they aim to activate property owners businesses, residents and the municipality.

The challenge then lies in selling the idea of sustainable development to business owners and municipalities, both of whom may be characterised by a desperate economic position. Politically and economically it is safer to follow the conventional development patterns. It is difficult to get beyond the perceived extra costs of environmental considerations as well as the risk factor of attempting a creative project. In the case of Toronto, the benefits of balancing environmental, social and economic issues is recognised conceptually in the Official Plan:

> Council shall promote energy conservation in order to conserve non-renewable resources for future generations and reduce pollution and long-term social costs. Council recognizes that energy conservation is a useful economic tool inasmuch as it can reduce costs for businesses and residents and provide jobs and new business opportunities. (Metropolitan Toronto, 1993, Section 2.22)

This statement leads me to question why energy conservation is not more strictly regulated by policy that directly affects action. The Commercial Facade Improvement Grant and Loan Program, for example, could integrate policies that ensure that the recipients plan to use energy efficient windows and lighting, as well as recycled building products. This would help to support environmental businesses while educating the main street businesses on their own impact. These types of policies and programs will bring the intent of the Official Plan into action by those who may not normally take the step.

From a private business owner's perspective, apathy towards sustainable development must be overcome by considering how a strategy can be developed that will meet their economic bottom-line. Focusing on the lucrative aspects of a sustainable street would involve creating a unique shopping experience only available along Yonge Street. It requires the generation of media attention and the resulting public attention. Generating activity takes more energy than business owners can provide on their own. This is one of the reasons why a range of community groups should be involved in the efforts. These groups can take some of the renewal burden off the shoulders of business owners while exemplifying that broader community will ultimately benefit from a healthy main street. In order to invite the public back to

Yonge Street, a dramatic change must occur to give them an incentive to change their normal shopping and lifestyle patterns. Further, by considering changes through an evolutionary process, the street can avoid a boom and bust process of continually planning one renewal project after the failure of another. Instead, the city can focus on a strategic plan that can grow and change to introduce new concepts and innovations in sustainable development.

Toronto's pre-eminent, sustainable main street will set the standard for other projects around the country. Due to the localised nature of this model strategy, the result will not lead to direct replications as in other renewal trends. Instead, this type of revitalisation project creates a unique experience by focusing on the local community. Over time, the sustainable street can support a new shift in the economy which takes into consideration full-cost accounting of consumer products, including their life-cycle environmental impact. The street as an awareness program can educate both businesses and consumers of their impacts and provide a new standard by which to conduct their transactions. Sustainable renewal projects are, therefore, a small but necessary step towards a stable urban environment. Overcoming the perception that economics and environmentalism are a dichotomy will transform sustainability into an opportunity instead of a barrier.

Note

1   J. Oosterveld is working with the private consulting firm, Lakeshore Planning Group in Toronto.

References

Barber, J. (1996) The mall is dead. Long live Yonge!, *The Globe and Mail*, July 25.

Barnes, W. Anderson (1986) Ghiardelli Square: Keeping a first, *Urban Land*, Vol. 45, No. 5.

Board of Health (1988) *Healthy Toronto 2000*, The City of Toronto, Toronto.

Brown, L.R. (1996) The Acceleration of History, in L. Starke (ed.) *State of the World 1996*, W.W. Norton and Company, New York.

Gibbs, D., and S. Braithwaite (1996) Moving towards sustainable development? Integrating Economic Development and the Environment in Local Authorities, *Journal of Environmental Planning and Management*, Vol. 39, No. 3.

Gorrie, P. (1991) New York's Instant Downtown, *Canadian Geographic*, Vol. 111, No. 2.

Gruen, N. (1986) Public/Private Projects: A Better Way for Downtowns, *Urban Land*, Vol. 45, No. 8.

Healthy City Toronto (1991) *Evaluating the Role of the Automobile: A municipal strategy*, City of Toronto, Toronto.

The Municipality of Metropolitan Toronto (1991) *Cityplan Proposals*, Toronto.

The Municipality of Metropolitan Toronto (1992) *The Livable Metropolis*, Draft Official Plan Document, Toronto.

The Municipality of Metropolitan Toronto (1993) *City of Toronto Official Plan Part I: Cityplan*, By-law 423-93, Approved July 20, 1993.

Nickerson, M. (1993) *Planning for Seven Generations: Guideposts for a sustainable future*, Voyageur Publishing, Hull, Quebec.

O'Connor, A. (1996) Undoing the Damage, *Vegetarian Times*, October.

Planning and Development Department (1995) *Downtown Yonge Street Community Improvement Plan*, City of Toronto, November 3.

Robertson, K. (1995) Downtown Redevelopment Strategies in the US: An end of the century assessment, *The Journal of the American Planning Association*, Vol. 61, No. 4.

Roseland, M. (1991) Toward Sustainable Cities, *Ecodecision*, December.

Sewell, J. (1996) Get ready for another Yonge Street clean-up try, *Now Magazine*, April 4-10.

Stokuis, J. (1984) Why can't downtown be more like a mall?, *Urban Land*, Vol. 43, No. 9.

The Yonge Street Business and Residents Association Inc (1996) *Newsletter*, Issue 3.

Toronto Planning Board (1971) *The Future of Toronto*, Toronto City Hall, Toronto.

Interviews

Danny Bellissimo, City of Toronto Economic Development Division Planner, interviewed November 4, 1996.

Bob Sniderman, owner of the Senator Restaurant, Yonge Street Business and Residents Association member, interviewed November 25, 1996.

Chapter 17

Urban Strategies for the 21st Century: Zuid-oost Industrial Terrain, Groningen

C. Moller[1]

17.1   Introduction

A very special atmosphere has been created in Groningen since the sixties revolutionary atmosphere. The intention has been to create a more meaningful and coherent society, especially in reaction to previous failures in post war European cities.

The quantitative growth of the 1960s in Groningen became an accumulation of private aspirations without any larger ambition and caused a social reaction. The social democratic collective ambition grew out of this reaction, focusing on qualitative concerns, and from this developed a larger idea or framework for cultural, economic and social elements of the city.

This included reducing the role of the car, enabling coherent space for other activities combined with cleaning public space in order to give it back its public meaning. Discussion about alternatives for worn out city elements, and the introduction of a pluriformity of uses and meanings into the centre became crucial to a dynamic view of the city and the need for change as a revitalisation agent. Reducing mobility and encouraging a diversity of service industries back into the inner city was a crucial achievement of this social democratic ambition.

Around 1980, the discussion on formal objectivity (creating symbols) in urban development was raised in professional circles, to develop the cultural identity of the city as a whole. The main spatial structure formed a fixed framework for continuously changing uses. These ideas were defined in the structureplan of 1987, and explored in a series of projects including the Verbindingskanaal area around the central station and new museum, and the new housing development at Hoornse Meer. These projects also explored a new type of public participation based upon the idea of planning by communication of which participation is an important part. This is initiated by a framework proposition for urban development which then encourages wide public involvement and comment with the final responsibility held by the city council to assess and define the outcome.

17.2   City Structureplan 'De Stad van Straks'

The structureplan for Groningen, 'De Stad van Straks' (Gemeente Groningen, 1994), has been developed within this cultural atmosphere. Through focusing on the idea of the inclusive city, the plan explores the need for new combinations and interaction of functions in strategic urban locations to operate as catalysts for sustainable revitalisation. This is explored in ways to encourage new spatial relations to develop a network of new interactions across the city and to allow opportunities of new connections between previously separated or mono-functional elements of the city. These become active lines of mobility within the city, which reach from inner-city areas to the outskirts.

17.3   Opening the Envelope

In the first half of 1995 the Municipality of Groningen initiated a series of forums and workshops on urban issues surrounding sustainability. The aim was to exchange ideas and propositions on urban questions for the 21st century and how these might relate to specific questions in Groningen. A diverse cross-section of professionals were invited who had an interest in the future shape of the city, together with a range of international speakers.

The framework of discussion for 'Opening the Envelope' (Moller, 1995) was developed through a sequence of three themes; Changing Global Conditions, Intelligent Urban Tools, and New Urban Strategies. The initial forum addressed the changing mechanisms of the global economy, global warming, issues of sustainable environmental engineering, and the changing use of time, to contemplate their positive and negative effects on urban environments. This debate set the scene for the crucial questions of how to address these dramatic changes, and what of our current urban practice could be useful in our rapidly changing society.

The second and third forums considered a series of new and existing urban design tools and strategies which if used intelligently might help us to deal with the changing global situation, and improve our understanding of how cities work and evolve.

17.4   Zuid-Oost: Case Study Project

It is within the atmosphere generated by the debates during 'Opening the Envelope' and the city structureplan 'De Stad van Straks' that particular projects within the city are now focusing on opening up the chances for a more compact, efficient, and ecological city. The city scale ambitions are being carried out at the scale of individual projects. The Zuid-oost industrial terrain is a case study project selected here to illustrate how these are being explored in an open and creative way together with the challenges of diversifying and strengthening the economic base and employment opportunities.

The Industrial Terrain Zuid-oost covers an area of 600 hectares to the southeast of the city. The area is traversed by motorway, rail and canal infrastructure, dividing it into complex, often disorientating elements. The transport infrastructure is often unclear with poor connections, forming restrictive barriers within the location.

Through these dislocations, Zuid-oost has become a barrier between the inner city and the agricultural landscape beyond.

Diversification of the economic base of the city has become a major priority with such high levels of unemployment in Groningen. The political ambition is for Zuid-oost to become a new engine for economic investment and new kinds of employment opportunities.

Framework Concept

The project has been developed on the concept of an open framework rather than a fixed plan. This is based on a series of three overlaid elements which are able to adapt to changes in environmental and economic conditions over long periods of time. These include:

1   *Lines of movement,* including cycle paths, motorways, roads, rail and water transport;
2   *Economic development locations,* including specific clusters and themes of particular industrial use (e.g., the Molenpark for organisations such as IKEA);
3   *Interactive Strips,* Urban and Landscape including structures, including the Hunze zone or Sontweg/Driebondweg which enable separate uses to overlap and interact in beneficial combinations.

The development framework is used as a game board to facilitate negotiation between differing interests including the minimum short and long term needs of; economic development, ecology, infrastructure, and social / cultural urban development.

Strategy

Utilising the impulse of the new key projects of 'Eemspoort' and 'Europapark' as attractors linking into the larger processes of the city, the strategy is to analyse, understand, and re-interpret the historic elements of Zuid-oost's existing natural, and industrial landscapes as qualities for future development. To re-evaluate their potentials through an approach of radical conservation, forging new meaning and inter-relations across previously separate elements, for example using the Hunze or the new Bremenweg. The strategy has been to create an open dialogue through a range of framework options to encourage input and participation from a wide variety of interest groups, businesses and inhabitants.

A constant and open negotiation between economic and environmental interests has been a key focus through out the project with an ambition to achieve the minimum interests of each while opening up potential and unforeseen inter-relations between their often conflicting needs, especially in the short term. These have included: to facilitate the equal importance to different interests in the design process; to ensure flexibility and adaptability in a rapidly changing and uncertain economic market; to achieve short-term requirements without limiting long-term potentials; and to keep opportunities open for future developments while not restricting short-term needs.

Process

The planning process has been structured to try and encourage interaction of different interest groups in a creative and open manner. The following steps have been followed towards this end: Phase 1, the development of an underlying flexible framework with input from a broad range of different disciplines, overlaid in separate layers of equal importance. This included the establishment of minimum 'starting conditions' from each discipline to achieve a 'Basic Packet'. Phase 2, the search for maximum potentials within each layer, developed from the minimum requirements of the Basic Packet. These were explored in environmental and economic scenarios. The intention is to open chances for combinations of elements between the two scenarios. Phase 3, The participatory involvement and reactions on the scenarios from politicians, interest groups, inhabitants and business people. Phase 4, the search for inter-relations between the two scenarios following public reaction and input. Phase 5, the development of a preferred scenario which seeks maximum possibilities between the alternative scenarios to develop the Structure Sketch. Phase 6, further discussion and reaction with politicians, inhabitants and the business community on the development of the Structure Sketch. Phase 7, the refinement of the Structure Sketch, into a comprehensive framework for development on site, and a phased development programme from 1996 to 2010 in three clear stages.

Transport Infrastructure

Accessibility throughout this part of the city has been largely defined by the historical development of largescale movement structures such as the major canals, motorways and railroads to Germany, the Randstad, and the Northern Region of the Netherlands. These infrastructures have encouraged industrial and commercial enterprise to locate here for ease of access directly to these networks.

Currently movement through the Zuid-oost industrial area requires restructuring and new investment especially for cycle networks which are almost non-existent .The strategy here is first of all for a clear understanding of the redefinition of the area through the development of new motorway structures which define new areas for urban development and the necessary infrastructure. A new railway station is planned in combination with a regional rail-road transferium to provide good intercity rail connections and a new city lightrail system which is to cross through the site. A comprehensive cycle network throughout the area will be realised at the same time with provision for good interchange points to public transport and clearly defined routes for social safety. All of these initiatives are intended to minimise reliance on private traffic while proposing major improvements public transport networks.

Economic Structure

The developing industrial sector in Zuid-oost is rapidly becoming the new economic engine for Groningen with two major new elements proposed. Eemspoort – a new distribution centre and Europapark – a new key location (on the old Powerstation site) are being redeveloped for Information and Telecommunications businesses. Both are connected directly to the main motorway and Europapark is also directly connected to the rail network through the construction of a new station and lightrail system. Between these two 'attractors' are a series of other themed areas for development including the Molenpark a new location for large warehouse retail activities, the developing Harbour area, the Milieuboulevard & Railterminal a new location for heavily polluting and recycling businesses, Driebond a developing auto boulevard. Each of these areas is being carefully located and phased in time to provide the maximum benefit to the further revitalisation of the existing business areas as well as to make the area more accessible and conducive to other activities such as offices, recreation and specific types of dwellings. In other words the area is being developed as a coherant part of the city rather than a removed and inaccessible industrial sector with no chance for complementary activities.

Landscape and Ecology

Groningen stands at the tip of the *Hondsrug* an historical sand ridge created by the last ice-age. The Hunze stream which runs through the Zuid-oost area defines the edge of this ridge. The stream has over time lost its meaning and value as a watercourse. Currently it is barely recognisable as a flowing stream, more as a series of disconnected irrigation channels.

Groningen has lately developed a city ecology framework plan '*De Levende Stad, een aanzet voor een stads ecologisch beleid*' (Veltstra, 1994). This document set out to provide a larger scale understanding of the overall patterns of various ecological structures within the city, including specific areas and strategies for improvement.

Zuid-oost has an important role to play within this larger framework. In particular the intention to connect innercity areas with the landscapes to the south-east of the city, and to provide a network of good quality east-west and north-south green and blue structures. In this respect Environmental Engineers Battle McCarthy were employed to develop early proposals (Battle McCarthy, 1995) as to useful approaches and techniques to be considered from the outset.

The water structures for the new development area include the separation of the ancient Hunze riverbed to enable an improvement of water quality in the area, which in turn should generate improvement of a series of different natural habitats along the riverbed. This will form an important new structure in the ecology network of the city, as well as a recreational green corridor for cyclists from the inner city to the countryside beyond. The existing agricultural water structures are to be used as the underlying structure and geometry for the new industrial plots as a practical and historically significant strategy for a new generation of economic and ecological development.

Energy

Thinking about industrial energy as a way to encourage greater efficiency, recycling and new methods of energy production was part of the underlying development strategy for Zuid-oost. To use energy as a tool to open up possibilities for employment opportunities and as an instrument to explore the potential for a greater diversity of mixed use in the area.

The six month research project to examine the viability of combined heat and power bio-mass systems (Battle McCarthy, 1996) has been further investigated for the Zuid-oost industrial area through the joint research of the Agricultural Economics Research Institute of the Netherlands, TNO Milieu, Energie en Procesinnovatie, and Battle McCarthy Environmental Engineers (Rijk et al., 1996). The research project has been looked at with the possibility of energy crops being developed as the first stage of land use transformation before being used as industrial land. Inter-relations between the larger agricultural region and urban development of Groningen can through such an energy strategy offer mutual benefits to both interests. For example, harvesting of short rotation coppice is an important boost for agricultural employment through more economic agricultural land-use during off-season periods. The Bio-Mass study shows that the approach is economically feasible and would be a valuable way for Groningen to reduce carbon dioxide emissions.

17.5 Conclusion

The research and initiatives that have been established for the Zuid-oost project are an important aspect of the overall sustainable strategy for urban development in Groningen and the region. It is through these initiatives that advances in knowledge can be applied and tested in the rigours of the planning process. The inter-relation between such research and the practical and economic requirements of real projects is crucial to the aims of achieving a long term sustainable landscape, as each has an important checking and balancing effect on the other. It is also from the partnership between economic and environmental concerns that an effective climate of innovation can be developed, encouraging openness to new ways of thinking and unforeseen possibilities.

Note

1   Chris Moller (B.Arch, RIBA, architect/senior urbanist) is currently working as senior urbanist for the City of Groningen. He worked as an architect in New Zealand before travelling to London in 1988 where he worked for Terry Farrell and Company on urban and architectural projects in the United Kingdom and Hong Kong. He is co-founder of the practice S.333 Studio for Architecture and Urbanism, with whom he has won several international urban design and architectural competitions, two winning entries of which are currently being developed. He is also currently working as senior urbanist for the City of Groningen. He has taught at the Academy of Architecture in Groningen and Tilburg, The Netherlands. Visiting critic at both Diploma and Post Graduate Housing and Urbanism Units, at the Architectural Association in London, England.

References

Battle McCarthy (1995) *Landscape Sustained by Ecology, An Ecological Landscape Structureplan for Zuid-Oost Groningen*, B.Mc. Publications, London.

Battle McCarthy (1996) *CIBOGA and Zuid-Oost, Feasibility Study for an Agricultural Fuel Energy System*, B.Mc. Publications, London.

Gemeente Groningen (1994) *De Stad van Straks, Groningen in 2005*, concept structureplan, Groningen.

Moller, C.D. (1995) *Opening the Envelope, Workshop Reports*, April – Changing Global Conditions, May – Intelligent Urban Tools, June – New Urban Strategies, Gemeente Groningen, Groningen.

Rijk, P., S. van Loo and R. Webb (1996) *Haalbaarheidsstudie van een biomassa gestookte warmte/kracht-installatie in de gemeente Groningen*, Landbouw Economisch Instituut, Publicatie 2.205, Wageningen.

Veldstra, W. (1994) *De Levende Stad, een aanzet voor een stads ecologisch beleid in Groningen*, Ruimtelijke Ordening, Gemeente Groningen, Groningen.

Chapter 18

The Right Place for the Environment: A Method for Area-oriented Environmental Policy

A. Schreuders and E. Hoeflaak[1]

18.1 Introduction

New York City is not like Yorkshire. The centre of Rotterdam is not a suburb. Every area has its own characteristics and qualities. The centre is alive and noisy. In a suburb you can hear the birds singing. *The Right Place for the Environment* links the environmental ambitions to the specific characteristics and potentials of an area. In this way the right environmental quality is put in the right place.

The Right Place for the Environment is the Rotterdam method for area-oriented environmental policy. This method answers two types of questions which environmental and urban planners encounter in every day Rotterdam practice.

Which Environmental Quality in Which Place?

Firstly there is a problem with regard to content: Which environmental aspects are relevant to a certain location in terms of densities, energy, sound or greenery? How do you weigh them up and which measures and standards belong with this? When looking for answers to these questions, the planner in a country like The Netherlands, with lots of activities in a small area, encounters the dilemma of the compact city: building in and onto a city at regional level means gain for the environment because vulnerable nature and open space are spared and because automobility is reduced. On the other hand, building in the city means at local level that there are high noise levels and soil and air pollution. Sustainability, quality of life and ecological quality appear to clash with each other. Environment versus the environment: a framework for making deliberations seems necessary. What do you count as being the most important in which cases?

A Clear Place for the Environment in the Spatial Planning Process

It is not always clear how the environmental aspects play a role in the spatial planning process: not for the planners, not for the council and sometimes not even for the environmental experts. What are the ambitions with regard to the environment? What influence do other variables that are filled in have on the quality of the environment? What is the right time for which considerations and decisions? How does the council put across clearly the choices with regard to the environment and spatial planning?

As an answer to the questions with regard to content and process, Rotterdam has developed a clear and consistent environmental approach for spatial planning: '*The Right Place for the Environment*', the method for area-oriented environmental policy.

The method takes as its starting point the structuring environmental aspects such as ecology and public transport infrastructure linked to land use. Other environmental qualities and measures at a certain location are adjusted to this. As a result, *The Right Place for the Environment* offers a solution for the dilemma of the compact city.

The method can furthermore provide a recognisable place for the environment in spatial planning. To achieve this it must be incorporated in the planning process in the correct way. The method should be used in an early stage of the planning process to formulate a clear environmental commitment. Alternatives can also be generated and choices finally explicitly discussed.

18.2   The Method

The Rotterdam method for area-oriented environmental policy consists of two steps. Step one focuses on distinguishing the various types of areas within a spatial plan. It brings structure into a planning area and provides ecology and sustainability (limitation of mobility). Step two works out the quality image that accompanies every type of area. This concerns spatial organisation, environmental quality and concrete measures. The emphasis lies on the quality of life.

Step One: Distinguishing types of areas

Distinguishing the types of areas occurs on the basis of the environmental aspects that have a structuring function. This is on the one hand the public transport structure (train and metro) and on the other hand the ecological infrastructure. Both networks have a long life cycle. Delft and Rotterdam have been connected with each other by a canal since the 13th century. Even today the canal is still an important structuring element. The same applies for the rail infrastructure, but also for the ecological structure. That is why ecology and public transport are the cornerstones of the method.

The method links intensity and differentiation of the land use to variations in quality of the public transport connection and nature: high-use intensity and a mixture of functions in well opened up locations, low intensities in locations with ecological quality. This approach creates eight types of area, which are illustrated in Table 18.1.

Table 18.1   Types of areas based on public transport connections and ecological quality

Public Transport connection	Ecological quality	Types of area	Qualities
very good –	basic	rail junction	High-use intensities, lots of mixing functions (offices/shops)
good –	"	public transport zone	Modal use intensities, mixture of residential/working (inc. businesses with high work intensity/ha)
moderate	"	car area	Low intensities, quiet residential area, public transport secondary, little mixing
"	basic	businesses/glass-houses/infrastructure	Place for uses that burden the environment, low intensities, public transport is secondary
"	– –	agricultural area/other greenery	Mono-functional agricultural area, low ecological value
"	increased	urban/recreational nature	Relatively high dynamic nature close to or in the city, primarily for recreation
"	– –	outside area	Agricultural area with natural qualities, transitory area, partly recreational use
moderate	high	natural area	Unique low dynamic natural area, protected

The first four types of area have a decreasing degree of public transport connection. The degree of urbanisation is adapted to this. In the area type 'rail junction', the public transport connection is very good. That is why high-use intensities (for example high residential/office density, many visitors) are desired here. In 'car areas', the public transport connection is secondary and therefore the densities are low. The last four types of area have an increasing ecological value. In agricultural areas the natural quality is relatively low, whilst it is of course first and foremost in natural areas.

Step Two: Qualities of area types

Each area type has its own broad quality image. This consists of specific environmental qualities or criteria which use the potential of an area to the full. For the various environmental aspects, the quality levels that go with each area type have been viewed.

The environmental qualities can be described in different ways. They can be described concrete, in the shape of standards, e.g. parking standards (varying from 0.7 to 1.3 parking spaces per residence) or water quality standards (varying from general quality to swimming water quality). For other aspects they can be described indicative e.g. in case of power supply (district heating or solar energy) or in case of the type of ecological connection (forest connection or river bank zone).

A number of examples are given below of the way in which environmental qualities are distinguished for area types.

Energy The area type 'rail junction' has different potential for energy saving measures than the area type 'car area'. In the area type 'rail junction', the high density means that a collective facility such as district heating is possible. In a 'car area' this is not possible and individual facilities have to be made such as active and passive solar energy.

Noise nuisance The area type 'public transport zone' has broad dispensation possibilities for road traffic noise (60 dB(A)), because within this area type mixing of functions (including businesses, shops and residences) is considered desirable along through roads. The area type 'car area' must have a quiet character and that is why the preferred limit value of 50 dB (A) is used.

Water quality As water quality and natural value are closely connected, in the area types 'rail junction' and 'public transport zone', which have a basic natural quality, basic demands have also been made with regard to the water quality. In natural areas the water quality should be high.

In tables, summaries have been made per area of what the desired quality levels are. An excerpt of one of these appears in Table 18.2 as an example.

Table 18.2 Excerpt from one of the tables with environmental qualities

Area type	Sustainability		Quality of life	Noise (dB(A))
	Density	Energy/Waste		
Rail junction	>60 res./ha >250 jobs/ha	• District heating • In-house or underground collection	Limited permissibility of burdening uses a. Category businesses b. Type of road c. Type of hotels and bars a. Category 2 companies b. Concentrations along trough roads, incidentally spread	Industry: 55 Motorway: 55 Local road: 60 Rail: 70
Public transport zone	35-60 res./ha >100 jobs/ha	• Poss. district heating • In-house or underground collection	a. Category 3 companies b. Roads in traffic areas c. Concentrations along trough roads, incidentally spread	Industry: 60 Motorway: 55 Local road: 65 Rail: 57
Car area	± 35 res./ha	• Solar collectors	a. Category 2 companies b. No trough roads in residential areas c. incidentally	Industry: 50 Motorway: 50 Local road: 50 Rail: 57
Business/ glasshouses	± 50 jobs/ha	If possible: waste island	a. Category (4) 5 companies, hazardous companies b. Motorways c. Incidental	N/A

Non-Area-Specific Environmental Aspects

Although for a large number of environmental aspects area-oriented differentiation is possible and desirable, there are naturally also environmental qualities which cannot be specified per area because 'they are needed everywhere and always'. Differentiation is in particular possible for aspects related to spatial structure, nuisance and quality of the surroundings. For environmental aspects which are directly related to health, a differentiation of standards is not acceptable. This means, for example, that for external safety and air quality no different standards apply for the various urban area types. Furthermore, for a number of aspects differentiation cannot be substantiated. For these aspects the same quality level is aimed for everywhere. Examples are energy-saving measures and sustainable construction.

18.3   Result: The Quality Image

When the desired quality level has been met for all the aspects, this results in an integrated quality image for each type of area. To give an impression of this a description of this kind follows for three area types.

Rail Junction

These are areas in town centres with a high degree of mixing of functions: shops, offices, residences and urban facilities. The intensities (e.g. residential density) are very high. Accessibility by public transport is of a high quality: an intercity or interregional station can be reached by bicycle and metro in little time. In view of the character of the area there are broad standards with regard to noise. The parking standard for residences is low. Energy supply occurs by means of district heating. Waste collection containers are under the ground. Car-free residential areas form oases in the beating heart of the city. A lot of attention has been paid to the quality of the external area and the maintenance is very intensive. Water has a primarily 'hard' form in this area: canals and sewers. The water has the functional quality level of 'fishing water'. Swifts and ducks can survive in this area.

Car Area

This is a quiet residential area with few other functions. The density is fairly low. The transport is primarily individual: car and bicycle, but there are also bus connections. The traffic function is subordinate to the residential function of the area. The parking standard per residence is high, but there are also areas which as large-scale semi car-free 'natural oases' attract specific target groups. There is a coarse cycle path network which connects the area with the city and the outside area. Rows of trees protect the area from energy loss. The power supply for each residence is above all organised individually: for example solar energy, both active (solar boilers) and passive (south-facing). The environmental pollution is very very low: in general the target values are met. The water has the form of ditches and lakes and has the functional quality of 'swimming water'. There are large parts with an increased natural quality where there are natural banks for example. The long-eared owl and marten are found.

Urban/Recreational Nature

In this area there is a mixture of urban and green functions. The area can be easily reached by bicycle, bus and car from the neighbouring areas; there are parking facilities on the edges and at concentration points, but there are also large semi car-free and quiet areas ('walking wood', cycle paths). The nature is relatively highly dynamic and aimed at fairly intensive recreation. The water has the functional quality of 'swimming water'. Newly constructed urban nature areas must be connected with nature areas in the immediate vicinity. The presence of carp, woodland birds and songbirds makes it clear that this area has an increased natural quality.

18.4   The Process: From Commitment – via Process – to Assessment

The planning process in The Netherlands is framed in a large number of procedures. To give the environment a clear place in the planning process, the method will be connected to the customary procedure of this process. In each planning process the following phases can be distinguished: initiation phase (determining the starting points), design phase (sketches, designs, choosing from variants, detailed design, elaboration) and decision-making phase (administrative determination). In each phase, *The Right Place for the Environment* plays a different role.

Initiation Phase: Environmental Commitment in Programme of Requirements

If the environment is to be given a clear place in the planning process, then the environmental commitment will have to be formulated at the start of the process. *The Right Place for the Environment* offers sufficient starting points to do this (see also 18.5). This environmental commitment can be incorporated in a programme of requirements. Then the environmental targets are clear for everyone. In order to avoid restricting the creativity and give the planners sufficient space for the deliberation process that is spatial planning, an environmental commitment will have to be formulated in the form of a range. For each aspect it thus is clear in advance what the basic quality is and what an optimum environmental quality yields. Preferably the programme of requirements is dealt with at the administrative level so that for everyone, council and planners, it is clear at an early stage what the ambition levels of a plan are.

Design Phase: Contribution to Variants and Choices

Between determining the ambitions and the final plan lies an unpredictable design process. In an iterative process choices are continuously made whilst working, step by step, towards the final design. The range formulated between the basic and target quality functions as an environmental compass in this. The challenge is to keep as close to the target quality as possible during the design process. The measures in *The Right Place for the Environment* can help when looking for alternatives and solutions. The application in the Nesselande plan (Figure 18.1) gives examples of this.

Note: Below, two variants for the Nesselande location have been included. On the left is the basic model with a metro station close to the Zevenhuizer Lake. As a result, locations with very high densities near to the lake are not within walking distance of the station. The station lies in the planned waterside park. On the right is an environment-friendly variant with the metro station more towards the north east, at the location of the water ('Bad'). As a result, the area with the highest densities is within walking distance of the metro station. Supralocal recreational facilities, neighbourhood facilities and the neighbourhood shopping centre remain close to the metro just as in the basic model.

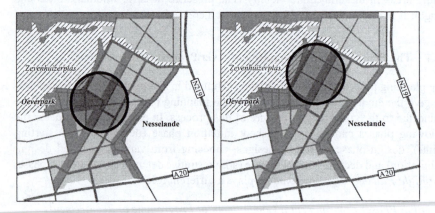

Figure 18.1   Using the method to look for environment-friendly variants

Decision-Making Phase: Assessment of Environmental Quality

At the end of the process, it will have to be clear for the council to what degree the plan brings environmental goals closer. *The Right Place for the Environment* offers a practical translation of the abstract and general environmental objectives into concrete qualities for individual spatial plans. Step one of the method (structuring in accordance with the public transport and ecological network) is the practical translation of the objectives in the field of mobility and ecology. Step two makes the connection with other environmental objectives such as water, energy and noise.

At the beginning of the process an environmental commitment is formulated on the basis of the method. By way of the criteria in the tables, this commitment creates a translation of general environmental objectives to a certain location. By comparing the environmental qualities achieved for each aspect with the range from the environmental commitment, it will immediately become clear what the environmental quality of the plan is. This shows to what degree the plan contributes to sustainable urban development. Figure 18.2 gives an example of how the administrative presentation can look.

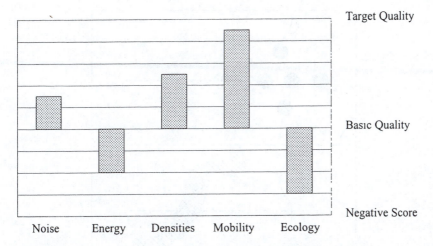

Figure 18.2 Example: presentation of achieved qualities

18.5 Making an Environmental Commitment

Planners can easily apply the method. The two steps of the method lead to an environmental commitment as follows: Step one relates to physical structure, cohesion and use. For this it is important to look further than the boundaries of the plan. Step two takes care of the spatial organisation and the desirable environmental qualities.

Step One: Structuring into area types

To assess the potentials within a planning area, a 'structure image' is made. Figure 18.3 shows how this can be done.

A. Urbanisation target picture On a map the current and planned public transport lines are drawn. Well connected areas are locations that lie within walking or cycling distance of a public transport station/stop. Urbanisation should occur in these areas.

B. Ecological target picture At the same time an ecological target picture is drawn up on the basis of the ecological qualities and potential. This only concerns the ecological quality of an area. Recreational use is subordinate.

C. Combination and trade-off A combination of maps A and B roughly indicates the structure of the desired development. Contradictions become visible. Deliberation occurs on the basis of two principles: 1. Ecology rather than urbanisation: even if a location is easily reached by public transport, nature has priority; 2. Compensation: the main rule can only be deviated from for green areas with a relatively low natural quality. But building on such a green area automatically means the obligation to create new nature with a similar size and quality in the immediate vicinity.

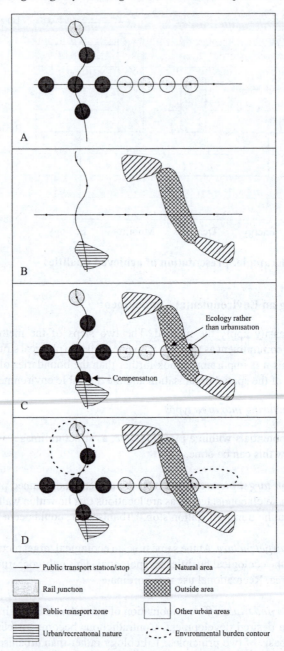

Figure 18.3   **Structuring into area types**

D. Test for the quality of life Finally, the 'structure image' now created is tested for its environmental burden. This above all applies to aspects of local environmental quality such as soilpollution, external safety and noise. If this test reveals any problems, a look will be taken to see whether they can be solved within the design, for example by adapting a function to the environmental burden observed.

The result of step one is the differentiation of the planning area into areas with different potentials.

Step Two: Environmental qualities for each area type

Each area type can be given environmental qualities with the help of the tables (see Table 18.2). In this way an environmental commitment can be arrived at. The criteria from the tables indicate the quality level that applies as the basic quality for that area type. Naturally local possibilities and limitations should be taken into account for this.

18.6   Scale Levels

The application of the method for a structure plan for the entire city is different than for a concrete project plan for 200 residences. At every level of scale different issues play a role. The general principles of the method are applicable at any level of scale. Urbanisation should be geared to the public transport infrastructure as much as possible and green structures should be protected. The elaboration of these general principles is different at every level scale. This means that some creativity is required of the planners when they apply it. The example below shows how the principle of urbanisation around public transport can be elaborated at various levels of scale.

For ecology it also applies that at structure plan level different elaboration is required than at lower levels of scale. At structure level a local park may not have any ecological importance. For a neighbourhood it can be of irreplaceable value. For planning at this level it will have to be taken into consideration as such. In case of contradicting elaborations at various levels of scale, the elaboration at the highest level will be leading. After all, it is at this level that the long-term structures are determined.

18.7   Other Choices

Application of the method can lead to other choices than were made in the past. An example will make this clear. For the sake of illustration, the method was applied in Rotterdam to the already realised Prinsenland zoning plan (Figures 18.5a and 18.5b). In Figure 18.5a the landuse and densities of the current Prinsenland zoning plan has been translated into the area types of this method, without using the method actually. Figure 18.5b shows the result of the application of *The Right Place for the Environment* for the same area.

Note: Due to its good connection to the public transport network, the entire new construction location of Schiehaven-Mullerpier can be typified as a public transport junction. The entire location is therefore suitable for residential densities of over 60 residences/ha. When a closer look is taken at location level, there is a clear difference between the individual sublocation close to the public transport stations/stops and the areas along the Nieuwe Maas. Any inhabitant in the first sublocations will be more likely to use the public transport. The locations along the water are on the edge of the sphere of influence of the stations/stops. Within the location the general principle of 'high densities around public transport' can again be applied. This yields a further differentiation in which the densities increase the closer you get to the public transport station/stop (increasing densities on the map from left to right).

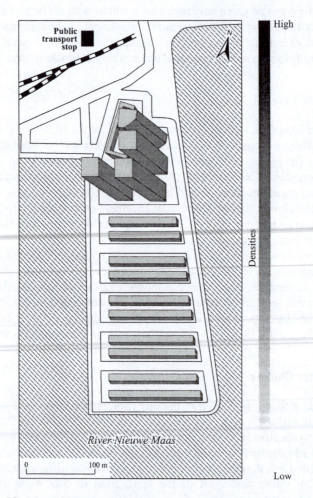

Figure 18.4   New housing construction versus public transport

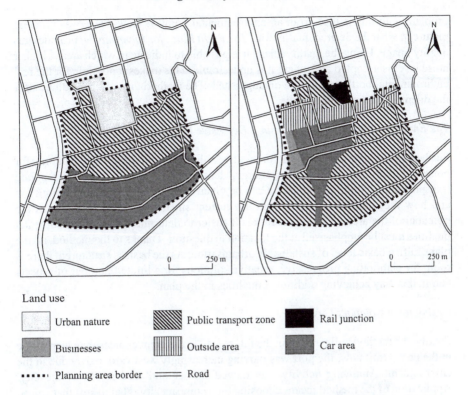

Land use

☐	Urban nature	▨	Public transport zone	■	Rail junction
▧	Businesses	▥	Outside area	▦	Car area
•••••	Planning area border	═══	Road		

**Figure 18.5a Land use according   Figure 18.5b Preferred land use
to zoning plan   according to the method**

Comparison of both figures shows a number of striking differences. The average density is approximately equal, but the differentiation is larger in the application of *The Right Place for the Environment*: it requires higher densities in a wide radius around the metro/express tram stops. Second, the location, function and organisation of the park is different in the area-oriented method. The green area has a connecting function between the Kralingsebos and the Schollebos woods and – in view of the character of both woods – a wet organisation. Finally, the Rotterdam method places a 'car area' in the heart of the planning area, where the densities are relatively low and the car ownership rate is relatively high. With an eye to the desired low noise burden (50 dB(A)), however, no through traffic is permitted in this part of the planning area.

18.8   Conclusions

Real Bottlenecks become Visible

It is clear that it is noisier or more lively in the city than in the countryside. The problem with this is that, from an environmental point of view, there was no clear picture of what the desired quality was. *The Right Place for Environment* provides this quality image. Using the quality image it is possible to indicate in which areas a higher noise level is accepted and where it is undesirable. This makes it possible for the real environmental bottlenecks to be distinguished. The method thus offers a solution for the dilemma of the compact city.

The Environment is Regarded not in a Defensive but in a Creative Way

Furthermore it is possible to look for creative solutions at an early stage. Not only for noise but, for example, also for energy, densities, ecological quality and public transport connections an indication can be given early on of what the ambitions are and how they can be achieved from the current situation. The method makes the potential of the area visible. This makes it easier to determine which environmental qualities must be emphasised at the location in question. Thanks to the method, all the plans fulfil a basic level of sustainable urbanisation. As the basis is structured, there is room for elaborating progressive, creative solutions for a limited number of aspects and in this way achieving additional qualities in the plan.

Opting for a Compact City

Finally, the method will perhaps in the future lead to different choices than were made in the past. Until now, the port was moving increasingly westward, partly due to the environment. Annoying activity was moved further and further out of the city. Application of the method means choosing for a compact city. That means that you no longer ask companies with a large disturbance area to set up in places where the fewest people live. This above all applies for companies with a large number of employees and/or visitors per hectare. You want to have them near to the urban area and public transport.

However, for companies with few employees per hectare and a high burden on the environment solutions will still be sought in areas with as low a population as possible. The method shows that high noise levels are a problem for sparsely populated areas (area type 'car area'). The consequence of this is that large-scale construction projects should not be carried out in these noisy locations.

Clear Process

With *The Right Place for the Environment* environmental ambitions can be formulated in an early stage. By already incorporating them in the initiation phase of a plan, the environment becomes an integral part of the design process, far more than is now the case. The methodical and therefore transferable approach in principle makes it possible for every planner to formulate the environmental commitment himself and incorporate it in his plan. What is more, the method makes the environmental quality clear for everyone by making use of scores for each aspect. A presentation in the form of bar charts will make it clear for the council at a glance what the environmental quality of the plan is. This makes it possible to present clear choices to the council.

Note

1   André Schreuders and Erik Hoeflaak are policy advisors at the Environmental Policy Department of the Municipality of Rotterdam in The Netherlands. They both advise the municipality on sustainable urban planning.

References

De Hoog ontwerp en onderzoek et al. (1995) *Strategie van Twee Netwerken*, R.P.D, Amsterdam.
Gemeentewerken Rotterdam (1997) *Milieu op z'n Plek, maatwerk voor milieu in ruimtelijke plannen*, Beleidsnota, Afdeling Milieubeleid, Rotterdam.
Hoeflaak, E., and E. Zinger (1996) Milieu op zijn plek, *Stedebouw & Ruimtelijke ordening*, No. 3.
Stad & Milieu (1995) *Waar vele willen zijn, is ook een weg*, Ministerie van VROM, The Hague.

Part E
Employing Indicators and Analysis to Achieve Integration

Design and Use of Urban Sustainability Indicators in Physical Planning: A View from Cascadia

D. Miller[1]

19.1 Introduction

City planning programmes in most mature industrial countries historically have focused on allocating space for urban activities, relating these to transportation needs and impacts, and providing desired infrastructure. In addition to improving the quality of life of residents, these plans have had economic growth and increased employment as major objectives. About 35 years ago, many of these planning programmes began to focus as well on social issues and especially on assessing needs and designing social service delivery systems.

More recently these planning programmes have begun to address as well environmental concerns in urban areas, both in terms of reducing pollution and in responding to the natural ecological dimensions of the urban site and its immediate hinterlands. Now the more forward-looking programmes are considering and experimenting with ways to integrate physical planning and environmental management. This was the principal theme of the first International Symposium on Urban Planning and Environment, sponsored by the Dutch Ministry of Housing, Physical Planning and Environment, and held in Seattle in 1994. As papers presented and discussed at that meeting point out, practices in physical planning and environmental management oftimes conflict (Miller and de Roo, 1997), but there has been considerable success in a growing number of instances to devise ways in which they can complement one another in an economically and politically feasible manner.

The next step in this progressive development is to integrate the social dimension with efforts to improve the environmental and economic conditions of cities (Beesley and Russwarm, 1989). Growth management acts adopted by a number of states in the U.S., for example, call for cities and counties to deal with these three sets of issues in plans which they are required to prepare. This combination of major concerns is also manifest in the growing urban sustainability movement. One demonstration that sustainability is becoming a central goals of planning in America is its use as the principal organising theme in the fourth edition of a major planning textbook (Kaiser et al., 1995).

While sustainability is increasingly accepted by planners and the public as a desirable ideal, there is considerable uncertainty over how to incorporate it into planning programmes. A promising approach is to specify social, environmental and economic objectives to provide guidance in developing and measuring the effectiveness of planning alternatives, and then to employ these measures in monitoring the implementation of these plans (Miller, 1997). My purpose in this chapter is to describe and assess five programmes in the central Puget Sound region which have developed and are employing urban sustainability indicators. One of these programmes is a voluntary effort by a non-governmental organisation and four are by local governments.

The Puget Sound urban region is located in the western portion of Washington State, is home to over three million people, and is one of the fastest growing areas in the US. It is at the center of a larger region, often referred to as Cascadia, stretching from Portland, Oregon to Vancouver, British Columbia in Canada. Some have called this larger region Ecotopia, both for its beautiful natural setting and because of the high level of popular support given to protecting environmental quality as rapid growth occurs.

19.2   What are Sustainability Indicators?

Urban sustainability indicators owe an intellectual debt to the social indicators movement, which sought to expand public and policy attention from the limited set of concerns addressed by economic indicators (Bauer, 1966). The development of social indicators raised issues of the purposes which they should serve, how ambitious they should be in covering a wide range of social concerns, and the criteria that should be used in their formulation (De Neufville, 1981; Carley, 1981). As Maclaren points out, urban sustainability indicators have been developed by a great many local and regional governmental units (1996), and are now the focus of several international efforts including the UN and the Sustainable Europe project by Friends of the Earth.

The term 'sustainability indicators' runs the risk of misunderstanding because each of these words mean different things to differing parties. Consequently, it is important to clarifying how this term is used in this investigation.

Sustainability

Sustainability refers to the long-term viability of human activity. Opschoor (1991) has defined sustainability as intemporally extended societal self-interest. On a somewhat more vernacular note, Herbert Stein has suggested that 'If something is unsustainable, it tends to stop'. This reminds one of the old adage, If your horse dies, we suggest you dismount. Each of these make the points that sustainability requires a longer-term view of phenomena than is popular, involves husbandry, and is of critical importance to the future of civilisation and of human survival.

Most discussions of sustainability include at least three dimensions of concern: people, place, and economy. The term *people* deals with the material and social well-being of all of the residents; *place* addresses ecological and other environmental concerns and commonly the quality of life which these support; and *economy* has to do with wise growth and development of employment in an increasingly changing and international market. Since sustainability requires viewing these with a long-term perspective, time is an important fourth dimension of most definitions.

While arguably no more or less important than the social and economic dimensions, environmental sustainability is perhaps less well understood and a source of major constraints to the other two. Environmental sustainability requires that a set of physical stocks and conditions be passed along to future populations that at least ensures economic and evolutionary development potentials comparable to current levels.

Sustainability raises several ethical issues. These include social equity: intergenerational, intergroup, and interregional. And these raise difficult political issues since they involve social redistribution of resources (Miller, 1985). Another ethical issue concerns whether the focus is anthropocentric or ecocentric. This too is a political issue with most decisions taking a human-centred view.

In this light, it is useful to observe that there are several sustainable futures. These differ as a function of different populations, including their incomes, and existing environmental qualities, along with other conditions. Since judgement is heavily involved in defining what constitutes sustainability, a practical planning approach would seem to be to pose several sustainable futures and to choose the acceptable ones through a political process.

Indicators

As with earlier economic and social indicators, those dealing with the three dimensions of sustainability are metrics with a purpose: measures with an ax to grind. Consequently they are normative in that they are techniques for scaling levels of performance or status of things held to be significant for sustainability. These measures provide a usually unambiguous means of assessing the 'distance' between current states and some reference or target situation. Similarly, indicators are useful in describing changes: movement toward or away from goals, the rate of this movement over time, and effectiveness in terms of degree of progress when related to the costs of a program. Thus the use of indicators implies evaluation such as comparison of a situation with a desired state, or with a past conditions including trends, or with another geographic situation.

Indicators are usually simplifications of the phenomena which they seek to measure, since many of these phenomena are complex and are often not fully understood (MOFA, 1982; Nagel, 1984). They thus represent a compromise between scientific accuracy and the demand for concise information to inform policy and guide decision making (Meyers, 1987). It is common that there is considerable available evidence concerning a topic of interest, but that this data is not useful to policy making because it is fragmentary, very detailed, not valid to decision making, and qualitative (Rutman, 1987).

A purposeful approach to formulating indicators suggests the value of considering the roles that they will be expected to play (Meyers, 1988; Cappon, 1990). These roles include accountability both in assessing options a priori and in monitoring an adopted course of action. Additional roles include shaping the agenda based on what has been selected for measurement, consciousness raising and constituency building as people become better informed about conditions through the use of indicators, and as a design tool for developing policies and programs.

This brief discussion of the forms, the nature, and the roles of indicators sets the stage for defining criteria to be used in designing useful indicators. As Maclaren (1996) points out, there is an extensive discussion of selection criteria and in the social, environmental, and sustainability indicators literature. As a summary and extension of this discussion, I suggest that eight criteria are of prime importance. These include:

- *Conceptually sound* – That indicators have a scientific basis: that they be reliable and desirably reflect cause-effect relationships.
- *Representative* – That indicators validly relate to targets: desirable or undesirable conditions.
- *Be quantifiable* – That indicators clearly represent the extent to which objectives are met or that conditions differ from targets, and be presented in physical units when this is appropriate.
- *Be attractive and expressive* – That the indicators be understandable and credible to the general public and to elected officials.
- *Address important concerns* – That the set of indicators has to be a manageable size in order to be effective: that they avoid the MEGO response, and aim for an AHA! response.
- *Robust* – That indicators be useful for measuring change over time, and across space, reducing the need for later respecification and duplicative data collection.
- *Relative ease of data collection* – That indicators be administratively feasible in terms of collecting scaleable evidence, and even employ secondary source data when this evidence is valid.
- *Avoid specifying the design of the action* – That indicators typify the conditions sought rather than the means to these ends, in a manner that prematurely presupposes solutions.

As is evident, several of these criteria relate to the practice concerns of planning, and in fact are based on lessons which I have learned from experience with planning evaluation (Miller, 1980) as well as with designing indicators for other purposes.

In summary, indicators are used for making comparisons, and the term targets has been used several times. The urban sustainability indicators projects which I will discuss use of several near-synonyms for targets, including: base line, reference point, benchmark, and critical level. Each of these suggest the desirability not only of measuring a feature of an existing situation, but also of specifying a condition with which this is compared. This point of comparison may take several forms. It can, of course, be a desired state based on objectives, adopted policy, or even standards (Miller and de Roo, 1996). Alternatively, it can be a past condition, either as something to be moved away from or something considered to be more desirable. Or the point of comparison may be situations found elsewhere, either in an overwise comparable area such as between similar cities or neighbourhoods, or with larger areas such as the region or nation. An analogy for these targets are the Plimsoll lines seen painted on ships, for use as reference points with the actual water level against the hull.

This discussion of sustainability and of indicators provides the basis for considering several urban sustainability indicator projects in the Seattle area. The next section of this chapter will describe these projects briefly, compare the indicators which are being used, address how these projects are related and how they differ both in content and in the procedures employed, and assess their relative effectiveness.

19.3   Urban Sustainability Indicators in the Central Puget Sound Region

The Pacific Northwest and British Columbia have been referred to as Ecotopia because of the exceptional natural environmental qualities of this region and because of the popular political support for conserving these qualities. Thus it is not a surprise that various units of government in this area have been among the first in developing indicators of urban sustainability. Among the first of these efforts were the Oregon Benchmarks and the British Columbia Sustainability Report.

Lately, the City of Seattle, King County in which Seattle is located, Pierce County to the south, and the Puget Sound Regional Council (PSRC) have all undertaken innovative programmes to develop and employ sustainability indicators. These efforts have been encouraged by provisions of the Washington State Growth Management Act of 1990 and 1991, which requires growth management plans at each of these jurisdictional levels, that policies on which these plans are based be co-ordinated among jurisdictions, and that urban development and the effectiveness of these plans be monitored. The deadline for initial adoption of these plans was 1994, and the third of these requirements resulted in elected officials and planning agencies looking to the new programmes in Oregon and British Columbia for useful ideas. The initial work of Sustainable Seattle, a non-governmental program, predated these governmental efforts, and has been an important additional source of inspiration.

Sustainable Seattle

This project began in 1990 as a grassroots effort to define what is valued in this community, and to devise a set of indicators to use in monitoring progress toward sustainability. These indicators address several quality dimensions, including cultural, economic, environmental, and social features of sustainability.

Sustainable Seattle is a broad-based citizen undertaking facilitated by the Metrocenter YMCA. While representatives from business, residents and citizens' groups, and public officials were initially invited to participate, the process has been an open one, involving over 250 people in development of the Indicators of

Sustainable Community report (Sustainable Seattle, 1995). The programme of Sustainable Seattle has been supported by contributions from a number of businesses and foundations located in the region.

In 1991, a diverse group of volunteers drafted a list of 29 potential key indicators and a set of secondary and 'provocative' indicators. In order to broaden participation, a Civic Panel process was initiated in 1992, including inviting over 300 citizen leaders from all walks of life to participate in a series of four workshops and on one of ten teams each dealing with a specific topic area. Based on discussion, feed-back surveys, and synthesis reports resulting from each workshop, the ten topic teams were each asked to develop and propose three to five indicators for their topical area. Many of these groups could only reach consensus by proposing a larger number. By Draft Version 3, there were a total of 99 indicators. At the fourth workshop, a year into this process, participants voted on these 99 indicators, using a green-dot-game in which each person could vote for 15 indicators. By the seventh draft iteration, a final set of 40 indicators was selected which would give a 'whole city' snapshop of movement towards or away from sustainability.

The first widely distributed publication from Sustainable Seattle was *Indicators of Sustainable Community – A Report to Citizens on Long-Term Trends in Our Community* (Sustainable Seattle, 1993). While this publication presented the master list of 40 indicators, it reported on research on 20 of these indicators and stated that research on the other 20 would be carried out over the following year. Each of the 20 indicators treated in this report were described and discussed in one or two pages, including time-trend data using the specified indicator, interpretation of the findings, an evaluation of the trend, and a discussion of linkages with other indicators. The results of the research on all 20 indicators were summarised in a 'report card' graphically showing for each of these whether there had been positive or negative change.

In 1995, a report with substantially the same format updated the 1993 publication and presented the results of research on all 40 of the sustainability indicators (Sustainable Seattle, 1995). Identification of these 40 indicators, and their classification under five categories, are shown in Table 19.1: the 1995 report card indicating the eight which have improved, the fourteen that have declined, and the eighteen which have remained unchanged.

The explicit intention in each of these reports is that they are the foundation for action, a means to the end of a more liveable and attractive future for Seattle. Initiatives to complement the indicators project include the Neighborhood Network of citizens who are being trained to be involved in planning at the neighbourhood level in Seattle which is currently underway, and in this capacity serve as advocates for sustainability as a major concern. A neighbourhood sustainability checklist has been developed to assist in this. Similarly, a *Seattle Guide to Sustainable Living* is being prepared for wide distribution to households, which will offer practical steps that individuals and families can take to further this goal.

Table 19.1   Indicators of sustainable community 1995

Declining Sustainability Trend (14)	No Discernible Trend or Unchanged (18)	Improving Sustainability Trend (18)
Environment		
Wild Salmon Wetlands Biodiversity Soil Erosion	Pedestrian-Friendly Streets Impervious Surfaces Open Space in Urban Villages	Air Quality
Population and Resources		

Residential Water
  Consumption
Farm Acreage
Renewable and Non-
  renewable Energy Use
	Vehicle Miles Travelled and
	Fuel Consumption
		Population
		Pollution Prevention and
		  Renewable Resource Use
		Solid Waste Generated and
		  Recycled

Economy		

Distribution of Personal
  Income
Health Care Expenditures
Housing Affordability Ratio
Children Living in Poverty
	Real Unemployment
	Work Required for Basic
	  Needs
	Community Capital
		Employment Concentration

Youth and Education		

Juvenile Crime	Adult Literacy
	High School Graduation
	Ethnic Diversity of Teachers
	Arts Instruction
	Volunteer Involvement in
	  Schools
	Youth Involvement in
	  Community Service	

Health and Community		

Childhood Asthma	Equity in Justice
	Low Birthweight Infants
	Gardening Activity
	Neighbourliness
	Public Participation in the
	  Arts
		Voter Participation
		Library and Community
		  Center Usage
		Perceived Quality of Life

Source: Sustainable Seattle (1995)

Table 19.2   King County performance monitoring indicators

Economic	Environment	Affordable	Land Use	Transportation
1. Employment in industries that export from the region.	9. Land Cover Changes over time; 1) percent impervious surface; 2) acres of selected cover types; and 3) changes in distribution.	21. Percent gap between the number of very-low, moderate, middle, and number of units affordable to them.	30. Percent of new dwelling units in a) Urban Centers, b) Urban Areas, c) Rural and Resource Areas.	41. Percentage of King County residents who commute one-way within 30 minutes between where they live and where they work.
2. Real wages per worker.	10. Number of days per year during which air quality is characterized as 'good', 'moderate', 'unhealthy', or 'very unhealthy'.	22. Percent of income paid for housing by very-low, low, and moderate income households.	31. Percent of new employment in a) Urban Centers, b) Manufactu-ring Centers, c) Urban Areas, d) Rural and Resource Areas	42. Ridership per capita per year as reported by METRO service sub-areas.
3. Real per-capita personal and household income in King County as a percent of the United States.	11. BTU consumption per capita, by energy type, per customer class.	23. Percent of persons and families who are homeless.	32. Percent of new residential units identified as re-development.	43. Percentage of County residents who travel daily by a) transit, b) HOV/Carpool, c) non-motor-ized, d) telecom-mute.
4. Percentage of population below the poverty level.	12. Vehicle miles traveled per year.	24. Percent gap between the price that the median income family can afford to pay for home and median price of affordable housing.	33. Ratio of percent increase in urban land consumption to percent increase in household and jobs.	44. Ability of goods and services to move efficiently and cost effectively through the region.
5. New business	13(a). Percent	25. Home-	34.Ratio of	45. Number of

	of ground water well samples that exceed Department of Health standards.	ownership rate.	actual density to allowed density to density of residential development.	lane miles of city, county, and state roads and bridges in need of repair and preservation.
created, net of losses (reported as number and percent of all businesses).				
6. New jobs created, net of losses, by employment sector (reported as number and percent of all jobs).	13(b). Change in 1) ambient monitoring conditions of Puget Sound and 2) water quality parameters of selected lakes and streams.	26. Annual rental vacancy rates.	35. Ratio of land capacity to 20-year household and job targets.	
7. Percentages of population over 25 with: 1) a high school diploma, 2) a tech. School degree, 3) an assoc. De-gree, 4) a BA/BS, and 5) a post-graduate degree.	14. Gallons of water consumed per capita and per customer class.	27. Annual rate of increase in average rent.	36.Land with six years of infrastructure capacity: a) Re-sidential, b) Commercial-Industrial.	
8. Percentage of 9th graders who go on to obtain a high school diploma.	15. Change in water levels of sample wells.	28(a1). Total public dollars per dollar of tax base spent for low income housing within each jurisdiction	37. Acres of urban open space and recreational land per thousand persons.	
	16. Change in wetland functions and acres by wetland class and drainage basin.	28(a2). Calcu-late total local dollars (inclu-ding in-kind value) per dollar of tax base spent for low-income housing within each jurisdiction.	38. Ratio of all jobs to all housing in King County and sub-regions.	
	17. Number of	28(b). Count	39. Acres of	

	blockages that interfere with continuity of Countywide habitat network.	number of units rehabilitated within each jurisdiction.	natural resource lands in a) forest land and b) farm land.	
	18. Change in number of wild stock and hatchery salmon per unit acre in indicator streams.	29. Count the number of units affordable to households earning 50% of median income within each jurisdiction.	40. Total number and average size of farms.	
	19. Rate of increase in noise from vehicles, planes, and yard equipment.			
	20. Tons of waste generated and recycled per capita.			

Source: GMPC (1995)

As a *Seattle Post-Intelligencer* article points out, the Sustainable Seattle indicators project has become a model for other cities, including San Francisco and London (Maier, 1995), and this project won an award at Habitat II in 1996. A similar project in South Puget Sound, by another voluntary organisation, has followed the lead of Sustainable Seattle in producing a similar report using twelve primary indicators (Sustainable Community Roundtable, 1993).

King County

In early 1995, the Metropolitan King County Growth Management Planning Council appointed the Benchmark Task Force consisting of elected officials from city and county governments and civic leaders. The task of this group, with the help of staff from three cities and the county, was to identify indicators which could be useful in monitoring and assessing progress with respect to Countywide Planning Policies (GMPC, 1995). As a point of departure, the Benchmark Task Force reviewed several sets of sustainability indicators, including those developed by Sustainable Seattle and the Oregon Benchmarks. The lead staff member for the Oregon Benchmarks Programme was retained as a consultant.

By spring of 1995, 90 draft indicators were proposed, reviewed, and amended at a Stakeholders Workshop. This list was reduced to 40 indicators, which were documented and taken to eight public workshops held around the county. Citizens were invited to add indicators and to prioritise the indicator list. The Benchmark Task Force reviewed the results of these workshops, and revised the draft indicator list. It also reaffirmed its earlier recommendation that targets be set for each indicator, as the 1996 phase of work in the benchmark work program, and that these indicators be published in the 1996 *King County Annual Growth Report.* The Growth Management Planning Council has agreed to review the benchmarks annually.

The resulting list of 45 performance monitoring indicators, classified under five categories, are summarised in Table 19.2. This set of indicators is intended to be the basis of measuring how well King County is doing as a people, place and economy: the three major dimensions of sustainability. Each indicator is discussed in terms of policy rationale, data sources, and geographic unit on which the data is to be reported.

The King County Benchmark Programme, unlike Sustainable Seattle, is the undertaking of a governmental unit. However it shares characteristics in seeking to balance technical quality with public involvement in identifying the topics which should be included and how these are measured. Both the Stakeholders Workshop and the eight public workshops were structured to inform participants and to solicit their recommendations, which were incorporated into the final list.

The multi-governmental (cities and county) collaboration in developing the Countywide Planning Policies is a requirement of the Washington State Growth Management Act. Multi-governmental involvement of both elected officials on the Benchmark Task Force and in its staffing are an outgrowth of this, and promises several benefits. Constituent units of government learn about sustainability indicators and their development, and are encouraged to develop a comparable set at the city government level. Similarly, citizen involvement in this process developed a political constituency for replicating this program locally. The Benchmark Task Force and its staff is working with the Seattle Commission on Children and Youth, and the King County Children and Family Commission, in developing a set of indicators on health and well-being, and on co-ordinating this effort with the county-wide planning indicators (KCCFC, 1995).

Pierce County

The Quality of Life Benchmark Project differs considerably from the King County indicators work. Table 19.3 shows a comparison of the number of indicators by subject matter for these and other programmes, and that there is some comparability on this basis. However, the Pierce County program employs 80 indicators, each of which address one of the three dimensions of sustainability (Pierce County, 1995).

Table 19.3   Categories and numbers of indicators for five Central Puget sound sustainability benchmark programs

	Jurisdiction and Number of Indicators per Category				
	Regional	*King*	*Pierce*	*Seattle*	*Sustainable*
Population	4	N/A	Note #1	N/A	Note #2
Economy	6	8	16	7	9
Housing	3	13	7	Note #3	Note #4
Environment	3	13	6	11	8
Land Use	N/A	11	12	7	Note #5
Transportation	34	5	8	Note #6	Note #7
Cost Effective Infrastructure	N/A	Note #8	6	N/A	N/A
Health and Safety (Health and Community)	N/A	Note #9	18	7	9
Population and Resources	N/A	N/A	N/A	N/A	7
Educational Excellence (Youth and Education)	N/A	Note #10	8	Note #11	7
Cultural and Recreation	N/A	Note #12	8	N/A	Note #13
Social Equity	N/A	Note #14	N/A	5	Note #15
TOTAL	51	50	89[Note #16]	37	40

Notes:
1   Two population indicators under 'proper distribution of land'.
2   Population indicators under 'population and resources'.
3   Four housing indicators under 'community' category.
4   One indicator under 'economy' category.
5   Measures farm acres, impervious land cover, and open space in urban villages.
6   Three transportation indicators under 'environmental stewardship' category.
7   Two indicators under 'population and resources' category.
8   One indicator under 'land use' (land with 6 years infrastructure capacity).
9   Working with Children and Family Commission to develop social indicators.
10   Two indicators under 'economic development' category.
11   Two indicators under 'economic opportunity and security' category.
12   Reports open space and recreational land.
13   Four indicators under 'youth and education' and 'community'.
14   Address social equity under 'housing' and 'economic development'.

15 Seven indicators under 'economy' and 'community'.
16 Since no historic data were available for nine of the indicators in the 'proper distribution of land category', they were not used in the composite done in 1995.

To make this larger set more manageable to interpretation, the 80 indicators are grouped into nine areas of concern, referred to as goal categories. Furthermore, the indicators within each of these categories are summarised by taking an unweighted average for each of the six years for which data was collected. This facilitates viewing change within each of these categories over time and, because the scale for each of these categories is constant, permits comparison between categories. However, these summary figures are necessarily not direct measures, but index numbers, which limits the ability of citizens to see how individual indicators are doing and to analyse the results.

Another interesting innovation of this project is the identification of four of the goals categories as being directly affected by public initiatives in the form of regulations and expenditures by public agencies. These four include education, infrastructure, land use, and public health and safety. The indicators for these categories are aggregated to show changes over the six time periods as well.

The Pierce County project notes that nine additional indicators dealing with proper distribution of land among three land uses are being developed, but because no historical data is available, they were not used in the current composite figures. Data for these will now be collected, permitting time series analysis in the future. While a small advisory committee assisted in developing the current set of indicators, the project intends to seek broader citizen input over the coming years.

City of Seattle

The growth management plan for Seattle, adopted in 1995, was titled *Toward a Sustainable Seattle,* and gained national recognition for its emphasis on developing a set of urban villages. This plan was guided by three major goals: environmental stewardship, social equity, and economic opportunity and security. These goals were developed through a series of workshops involving over 1200 citizens, in formulating a set of framework policies. However the plan itself was prepared with little direct citizen input which contributed to considerable controversy and organised opposition (Miller and Holt-Jensen, 1997).

This controversy resulted in a reappraisal of the planning program and an active public involvement process for neighbourhood planning to refine the comprehensive plan. It is thus surprising that little effort has been made to secure public involvement in developing the set of indicators which will be used in monitoring and evaluating the performance of the comprehensive plan.

This project has identified 37 indicators (Table 19.3). Twenty-one of these are also indicators included in the King County Benchmark programme. It is the stated intention to work with the county benchmark staff to assure that the data are comparable, which will facilitate comparisons of these indicators (Seattle, 1995). A number of these indicators are also found in the Sustainable Seattle reports, confirming the influence of that programme on the City of Seattle Indicators Project.

The 37 indicators for Seattle are classified under five categories including, in addition to the three major goals developed during the framework policies process, community, and the urban village strategy. This list was developed by the planning staff, discussed with the city council, revised, and adopted by the council in 1995.

Puget Sound Regional Council

The PSRC has responsibility for regionwide land use and transportation policy under the Washington Growth Management Act. Its region consists of the four counties making up the central Puget Sound metropolitan area: King, Pierce, Kitsap and Snohomish counties.

The Regional Council has prepared and adopted a regional plan: Vision 2020. In revising this plan, they developed a set of policies and are in the process of designing indicators to use in implementation and performance monitoring. Used in implementation monitoring, the indicators are intended to answer the question: Did we do what we said we were going to do? In performance monitoring, the question is: Does it appear that our actions have produced the desired outcomes?

The Council staff have formulated 51 indicators, almost all of which address aspects of sustainable development (Table 19.3). Many of these indicators also appear in the set adopted as part of the King County Benchmark programme, and a number are adopted from the work of Sustainable Seattle. However the Regional Council's indicators take a narrower view than do these other programs, with most of the emphasis on housing, land use, population, and especially transportation which accounts for 34 of the indicators (Table 19.4). These represent the mandate of the Regional Council.

The Regional Council is committed to use these indicators in several of its publications: several technical reports, a benchmark report, and a regional report card. While its indicators project is the last to get started of the five reviewed here, this has facilitated incorporation of lessons learned from the other efforts and positions the Council to act in a co-ordinative capacity as these other sets of indicators are revised and refined over the coming years. This sequence in the development of programmes has also resulted in greater understanding and support by the elected officials of local governments serving on the Regional Council for undertaking this indicators programme.

19.4   Comparison of the Five Sustainability Indicators Programmes

As the previous description of the four governmental and one non-governmental indicators programmes reveal, there are both similarities and dissimilarities between them. These five cases thus provide a range of approaches to developing indicators, several roles that they are intended to serve, variation in how they are presented, and differences in the number and content to the indicators.

Intellectual and Political Debts of the Programmes

As was pointed out earlier, the Sustainable Seattle programme preceeded the others and is explicitly credited as a source of ideas by most of the other studies. Sustainable Seattle has developed a public constituency for the use of sustainability indicators which made it politically attractive for local governmental units to adopt this strategy and provided a starting point for these other efforts. This is reflected in the Vision 2020 award recently presented by the PSRC to Sustainable Seattle.

The King County Benchmark programme not only drew on the work of Sustainable Seattle, but makes extensive reference to the work of the Oregon Benchmarks Programme. As noted previously, both Seattle and the PSRC used the Sustainable Seattle and King County indicators sets as sources of ideas.

In short, these programmes are highly interrelated. This results in considerable comparability between these sets of indicators. At least equally important, the earliest of these programs paved the way politically for initiating subsequent programs. Elected officials became familiar with this approach and saw the political support for and advantage in establishing complementary programmes. The provision for monitoring in the Washington Growth Management Act was an added incentive to adopt this approach.

Procedures Employed in Developing Indicators

Sustainable Seattle is a voluntary, grassroots effort to develop indicators, as a bottom-up political strategy for educating the broader public to existing conditions and their change over time, and to influence public policy and programs to serve the multiple goals of sustainability. The King County program began as a technical exercise under the direction of the Benchmark Task Force consisting of elected officials from several jurisdictions and representatives from the community. Early in this program, a wider group of stakeholders were involved, and drafts were then taken to a number of open public workshops, which resulted in changes in the final recommendations. Considerable attention was given to learning from the public and building a constituency for the program and its results.

The Pierce County and Seattle programmes, in contrast, have been largely technical exercises with the intention that future revisions will involve more public participation. In both the Seattle and PSRC cases, it can be argued that wide public participation in the King County programme serves to validate at least those indicators which were adopted from the county program.

Content of the Sets of Indicators

As Table 19.4 shows, both the PSRC and King County programmes are setting up targets against which to compare current levels of performance as measured by the indicators. Establishing targets is an exercise in judgement, consequently will need to involve a political process, and may raise controversy.

Table 19.4 General comparison of five urban sustainability benchmark programs

Jurisdiction	Scope	Geography	Definitions			Use of Targets	Level of Analysis	Indicators Summary
			Indicators (3)	Target (4)	Benchmarking (5)			
Regional Council	narrow (1)	centers, UGA, rural, 7 sub-areas	agree with below definition	agree with below definition	agree with below definition	currently working on targets	include description	yes
King County	broad (2)	County, 3 sub-areas	agree with below definition	agree with below definition	agree with below definition	currently working on targets	?	?
Pierce County	broad (2)	County	agree with below definition	agree with below definition	agree with below definition	no	include description	yes
Seattle	broad (2)	Seattle, 5 centers, 25 urban villages	agree with below definition	agree with below definition	agree with below definition	future	?	?
Sustainable Seattle	broad (2)	King County, Seattle	agree with below definition	agree with below definition	agree with below definition (6)	no	include description	yes

1   Focus on growth and transportation indicators.
2   Include social indicators and more detail.
3   Indicator – 'Key performance measures for which quantifiable or directional targets may be set' (*VISION 2020, 1995 Update*), 'A measure that reflects the status of a system. Examples: the Dow Jones Industrial Average, the number of spotted owls in a forest ecosystem, and oil pressure on an engine' (*Sustainable Seattle*).
4   Target – 'A numeric goal or stated direction to be achieved that reflects the policy commitments of VISION 2020' (*VISION 2020, 1995*).
5   Benchmarking – 'Benchmarking involves a two-step process: identifying key performance indicators that highlight the effects of steps taken by the region to implement VISION 2020, and establishing corresponding prformance targets, or benchmarks, that the region hopes to achieve over time' (*Regional View, June 1994*).
6   Sustainable Seattle defines a benchmark as 'a point of reference or a standard against which measurements can be compared; sometimes a goal or a target. Examples: Record highs in the stock market, optimal water levels in wetland, so-called 'full-employment' levels of acceptable unemployment'.

Each of the five programmes include a differing number of indicators, with differing substantive content, as shown by Table 19.3. This is in part a function of the mandate for each of the governmental units. Of the 51 indicators used by the PSRC, for example, 34 focus on transportation issues, reflecting the role of this agency in dealing with regional transportation. As another example, seven of the 37 indicators adopted by Seattle address the urban village strategy, which is the major feature of the new comprehensive plan. Even so, a number of indicators are common to most and in some cases all of these programmes, and the prospect is good that there will be more commonality as the result of future revisions where comparability can be shown as being valuable.

In terms of the five frameworks which Maclaren (1996) presents, each of these five programs appear to employ what she calls 'domains' as an organising device. While she prefers a causal framework, I argue that this is simply a programmatic application of indicators, especially if measures of inputs are avoided as is predominately the case with the five programmes reviewed here. Discussion of causal relations can take place in the exposition accompanying each indicator as, for example, the Sustainable Seattle publications do and as is found in the technical documentation for most of the other four programmes. This places less burden on having accurate causal knowledge, which is often in short supply or controversial.

The use of domains as an approach and organising principle in three of these programmes relates to their role as providing benchmarks for monitoring the progress of working programmes with respect to plans. The term benchmarks implies that the indicators are measures of progress, and appears to have resulted in broader public support than might have been the case if they had been called sustainability indicators, although the content is largely the same.

Presentation of Indicators and Report Cards

Each of these programmes have or intend to issue reports which provide an overview of how well the particular jurisdiction is doing in moving toward sustainability. These report cards will offer the public an opportunity to review not only how great various problems continue to be, but the effectiveness of programmes in addressing these problems.

As Table 19.4 shows, the PSRC, Pierce County, and Sustainable Seattle programmes seek to summarise performance over several indicators, commonly the several indicators within a topical category. In the Pierce County case, this is done by taking an unweighted average over subgroups of indicators. This facilitates constructing bar graphs of performance for each group of indicators for each of six recent years, providing information on changes over time. This programme recognises that an unweighted average assumes that all of the indicators and the conditions which they measure are of equal importance. Developing a differentiated set of weights for the indicators, or for categories of indicators, will likely be more controversial, is being considered as a next step in refinement of this programme.

The Sustainable Seattle programme offers a summary of performance over all indicators and for each of the five topical categories. It does this by using pie charts showing the proportion of indicators in each group which are trending away from or toward sustainability, or for which there is discernible trend. While this provides a general indication of relative improvement by category, it conveys no information concerning the magnitudes of these improvements since it employs nominal level data.

Intended Roles of the Indicators

These indicator programmes differ in their stated purposes. The Pierce County and Sustainable Seattle programmes are intended to measure and convey changes which are occurring over time, for public information and provide the information basis for advocacy. The King County, Seattle, and PSRC programmes are primarily intended to facilitate monitoring the effectiveness of planning policies and programmes.

The PSRC programme, for example, makes the point that these indicators will be useful in measuring their actions and the effectiveness of these, thus improving accountability. The Pierce County programme aggregated and summarised those categories especially subject to public policies and programmes, as a means of tracking the effectiveness of these efforts.

As the general discussion early in this chapter pointed out, indicators are measurements made for a purpose. Without an intended role, such measurements are simply data. Consequently, the selection of topics to address and the design of indicators for these necessarily depend on the role that they are expected to play. This is demonstrated by the variability in these five programmes.

19.5 Conclusions

The five urban sustainability indicators programmes reviewed differ in a number of ways even though they are interrelated in their development. These differences help them to serve as models useful in designing improved approaches to developing and presenting such indicators.

One of the primary differences among these programmes is the extent to which public participation played a role in their development. Especially in the cases of Sustainable Seattle and King County, this participation was considered to be of critical importance, both in identifying and attaching priorities to the indicators. Technical rigor was sought, but accurately reflecting the concerns of citizens and securing their support was paramount. Consequently, these two projects employed a bottom-up, mobilising strategy rather than a largely technocratic strategy.

Each of these five programmes benefited by having a political constituency. These constituencies are a function of the wide public support in the Northwest for sustainability and concern for the quality of life which depends on successfully meeting the three major goals of sustainability. These constituencies are also a function of political engagement, the desire and sense of right that citizens should have a say over matters which affect them, and consequently a demand for accountability both from plans and elected officials.

Additionally, several of these programmes built on the lessons learned in developing earlier programmes. This has resulted in improvements in identifying and specifying indicators, and has encouraged elected officials to countenance and even promote the development of sets of indicators. While somewhat conjectural, the evidence from these five programmes suggests that we are developing a culture of evaluation in the central Puget Sound region: a number of governmental units and a citizenry knowledgeable about indicators and insistent that the progress of policies, plans, and programmes be measurable and thus accountable to the public.

Notes

1   Donald Miller is Professor in the Department of Urban Design and Planning, University of Washington, Seattle. His teaching and research deal with planning analysis and forecasting, evaluation, the future of cities and of planning, urban spatial structure, planning theory, strategies for improving urban environmental quality, and the politics of planning.

2   The author wishes to thank the following people who supplied documentation and answered questions about their respective programs: Kara Palmer, Sustainable Seattle; Cynthia Moffitt, King County; Tom Hauger, City of Seattle, and Dr. Norman Abbott, Puget Sound Regional Council.

References

Bauer, R. (ed.) (1966) *Social Indicators*, M.I.T. Press, Cambridge, MA.

Beesley, K., and D. Russwarm (1989) Social Indicators and Quality of Life Research: Towards Synthesis, *Environments*, Vol. 20, No. 1.

Cappon, D. (1990) Indicators for a Healthy City, *Environmental Management and Health*, Vol. 1, No. 1.

Carley, M. (1981) *Social Measurements and Social Indicators*, Allen and Unwin, Worchester, MA.

De Neufville, J. (1981) Social Indicators, in M. Olsen and S. Micklin (eds) *Handbook of*

Applied Sociology – Frontiers of Contemporary Research, Praeger Publishers, New York.

GMPC (Metropolitan King County Growth Management Planning Council) (1995) *Benchmark Task Force Report*, GMPC, Seattle, WA.

Kaiser, E., D. Godschalk and F. Chapin (1995) *Urban Land Use Planning*, Fourth Edition, University of Illinois Press, Urbana and Chicago, IL.

KCCFC (King County Children and Family Commission) (1995) *Healthy Children, Youth and Families in King County*, KCCFC, Robert Wood Johnson Foundation Seattle Child Health Initiative, Seattle Commission on Children and Youth, Seattle, WA.

Maclaren, V. (1996) Urban Sustainability Reporting, *Journal of the American Planning Association*, Vol. 62, No. 2 (Spring).

MacRae, D. (1985) *Policy Indicators: Links Between Social Science and Public Debate*, University of North Carolina, Chapel Hill, NC.

Maier, S. (1995) Seattle's Long-Term Health is on the Wane, Survey Shows, *Seattle P-I*, November 16, B 1-2.

Meyers, D. (1987) Community Relevant Measurement of Quality of Life: A Focus on Local Trends, *Urban Affairs Quarterly*, Vol. 23.

Meyers, D. (1988) Building Knowledge About Quality of Life for Urban Planning, *Journal of the American Planning Association*, Vol. 54, No. 3 (Summer).

MFOA (Municipal Finance Officers Association) (1982) *Indicators of Urban Condition*, Government Finance Research Center, Washington, DC.

Miller, D. (1980) Project Location Analysis Using the Goals Achievement Method of Evaluation, *Journal of the American Planning Association*, Vol. 46, No. 2.

Miller, D. (1985) Equity and Efficiency Effects of Investment Decisions: Multicriteria Methods for Assessing Distributional Implications, in A. Faludi and H. Voogd (eds) *Evaluation of Complex Policy Problems*, Delftsche Uitgevers Maatschappij B.V., Delft.

Miller, D. (1997) Dutch Integrated Environmental Zoning: A Comprehensive Program for Dealing With Environmental Spillovers, in D. Miller and G. de Roo (eds) *Urban Environmental Planning*, Avebury, Aldershot.

Miller, D., and G. de Roo (1996) Integrated Environmental Zoning - An Innovative Dutch Approach to Measuring and Managing Environmental Spillovers in Urban Regions, *Journal of the American Planning Association*, Vol. 62, No. 3 (Summer).

Miller, D., and G. de Roo (eds) (1997) *Urban Environmental Planning – Policies, Instruments and Methods in an International Perspective*, Avebury, Aldershot.

Miller, D., and A. Holt-Jensen (1997) Bergen, Norway and Seattle, USA: A Tale of Strategic Planning in Two Cities, *European Planning Studies*, Vol. 5, No. 2.

Nagel, S. (1984) *Public Policy Goals, Means, and Methods*, St. Martins Press, New York.

Opschoor, H. (1991) GNP and Sustainable Income Measures: Some Problems and a Way Out, in O. Kuik and H. Verbruggen (eds) *In Search of Indicators of Sustainable Development*, Kluwer Academic Publishers, Boston.

Pierce County (1995) *Pierce County Quality of Life Benchmarks*, Pierce County Department of Community Services, Tacoma, WA.

Rutman, L. (1977) Planning an Evaluation Study, in L. Rutman, *Evaluation Research Methods: A Basic Guide*, Sage, Beverly Hills.

Seattle, City of (1995) *Comprehensive Plan Monitoring and Evaluation – Revised Preliminary List of Indicators*, City of Seattle Office of Management and Planning, Seattle, WA.

Seattle Planning Department (1994) *The Mayor's Proposed Comprehensive Plan, 'Toward A Sustainable Seattle'*, Seattle Planning Department, Seattle, WA.

Sustainable Community Roundtable (1993) *State of the Community*, Sustainable Community Roundtable, Olympia, WA.

Sustainable Seattle (1993) *Indicators of Sustainable Community – A Report to Citizens on Long-Term Trends in Our Community*, Sustainable Seattle, Seattle, WA.

Sustainable Seattle (1995) *Indicators of Sustainable Community – A Status Report on Long-Term Cultural, Economic, and Environmental Health*, Sustainable Seattle, Seattle, WA.

Chapter 20

A Case for Local Environmental Indicators: Participation in Quito, Ecuador

S.D. Vásconez[1]

20.1 Introduction

A well known issue in urban environmental planning processes is stakeholder participation. Evaluating past and present experiences it has become evident that active public participation in development strategies has proven to be more efficient and effective than top-down approaches. Due to its positive implications, participatory mechanisms have increasingly been incorporated in the design and implementation of planning strategies. Presently, the trend shows that central and local urban agencies world-wide are exercising a diversity of such methodologies. One of these mechanisms are environmental indicators generated at the local level.

Environmental indicators generated at the local level can contribute to the discussion by offering reference points or clues on where environmental quality stands in a determined territory. The formulation and implementation of indicators result from efforts to provide a clear picture of the environmental situation in a given area. These efforts derive their strength from the fact that aggregate indicators such as GNP or income per capita are not specific enough to give clues on the environmental trends experienced at intermediate and micro levels.

Rather than providing an in depth justification for participatory mechanisms at the urban environmental intersection, the discussion will address a particular proposal developed by the Metropolitan Municipality in Quito, Ecuador to abate environmental and social impacts caused by the severe contamination of one of the city's main river courses, the Machángara. In this proposal, local environmental indicators were the mechanism chosen because of the urgency of the conflict. As it stands, integrated solutions to the river's contamination will only be achieved in a medium to long term planning horizon, without consideration of their present effects.

In addition, local indicators methodology was conceived as feasible because it does not require large investments, and as similar experiences have attested (i.e., Seattle Sustainability Indicators and Oregon Benchmarks for Sustainability) can contribute to fostering a co-operative atmosphere between the stakeholders and local officials. Also, this strategy was envisioned as an initial step to promote a greater environmental awareness – an education process – that would eventually generate a transformation in the population's attitude and behaviour towards the environment.

As a starting point, a general framework for environmental indicators will be outlined. Attention will focus on the definition of environmental indicators in the local context. Under the methodological framework 'Pressure-State-Response', inclusion of participatory avenues will be discussed. In the next section, Quito's urban-environmental configuration will be discussed within the context of the larger national urbanisation tendencies. Finally, the third section will focus on a concrete proposal pursued by the Municipality to mitigate the increasing impacts of water contamination using local environmental indicators.

20.2   Local Environmental Indicators

Environmental indicators can be defined as quantitative and qualitative instruments that measure environmental conditions in a given territory. Because they are not fixed standards they operate as guiding tools for communicating information and promoting action (Hammond et al., 1995). They are also key for the simplification of environmental information and contribute to: quantify environmental conflicts and problems; analyse environmental relations and the impacts these have; communicate and transmit information about the environmental conditions in the locality; assess and define public priorities in relation to environmental quality; and formulate and establish action strategies for the solution of environmental problems.

Local environmental indicators strategies constitute an example of bottom-up approaches which can lead to effective negotiation and resolution of urban environmental problems. Their main goal is to guide policy makers' actions to address and solve environmental conflicts. In the urban context, indicators have great relevance to reveal the specific urban-environmental dynamics. Since indicators are not fixed measurements, but rather dynamic estimates, they help in collecting information as well as plotting trends. Their legitimacy lies in that they can assess the advancement or deterioration in the environmental quality of a given territory, contributing in efforts to monitor and control prior agreed standards.

In constructing environmental indicators, a framework that has proven to be very effective is the 'Pressure-State and Response' model (Hammond et al., 1995). This framework begins by asking a simple set of questions about the state and characteristics of the environmental unit to guide the construction of indicators: What is happening with the environmental resources? Why is it happening? What are the effects of these interactions? and What are the effects or responses elicited by these problems? At the local level, the Pressure-State and Response framework operates quite effectively since it can be easily employed with participatory mechanisms.

The 'Pressure-State-Response' model encourages the population to reflect about the current and future state of the environment in their locality. First, 'pressure' indicators reveal the agents (present or potential) causing environmental degradation. In turn, these portray a certain environmental situation or state. Following the environmental prognosis, the 'response' indicators focus on what can be done to deter or mitigate the present situation. These responses can be implemented in a short time span or in the long-term future.

When applying the 'Pressure-State-Response' model at local levels, especially when using communal participation, it is necessary to keep in mind the extent of the information to be collected. For this it is useful to formulate a scale for measuring environmental impacts. This scale will contribute to determining priority areas according to levels and degrees of effect.

A first step in participatory mechanisms is to organise an initial evaluation with the stakeholders. Local actors are the best available source for generating a comprehensive picture of the environmental situation in the area. Therefore, to take advantage of their expertise, it is necessary to channel their information through the organisation of a series of public discussions, meetings or focus group interviews (e.g. Delphi methodology). Once base line information is obtained, qualitative indicators generated by the community should be contrasted with quantitative indicators and standards. It is here that the role of participation becomes a great asset for the effectiveness of planning instruments.

In order to be effective, quantitative indicators need to have resonance in the community. If they lack this characteristic, translation of technical information to the communal level is not possible. People have to adopt indicators as their own, if these are going to be legitimate and valid. By contrasting the quantitative and technical environmental information steps need to be taken to incorporate in a systematic way the 'qualitative' perceptions held by the local actors.

20.3   Quito and the National Urbanisation Process

Similar to other cities in Latin America the urban-environmental configuration of Quito, as well as in other large settlements in Ecuador such as Guayaquil and Cuenca, has resulted from of a combination of complex economic, social, political and cultural forces. These demographic and population pressures were triggered in the 1960s, with the implementation of import substitution development strategies that lead to a grave crisis in the agricultural sector. The result was a massive rural migration to the urban centers (Carrión, 1987b; Larrea and García, 1995).

By the end of the 1970s, while maintaining a very high population growth rate, Ecuador had became one of the fastest urbanising countries in the region (Burgwal, 1993). In this context, Quito and Guayaquil served as the main refuge for migrants, who in search for better life opportunities found themselves incorporated into a deeply segregated urban scenario.

In response to the accelerating urbanisation process, during the late 1960s the government created planning offices in most municipalities. However, few of these offices developed comprehensive urban management policies and none were able to furnish integrated economic, social and environmental development policies (Hoshino, 1994). Many of the planning efforts of the past decades remained as good will documents with little or no implementation. Currently, only the metropolitan settlements of Guayaquil and Quito and perhaps Cuenca, are advancing to formulate integrative urban policies which place environmental quality as a major variable. The rest of the municipalities are still caught up in infrastructure and basic needs provision, which although greatly enhanced since the 1970s is still not universal. For example, by 1995, only 84 per cent of the total urban population had access to potable water and 70 per cent to proper sanitation (UNFPA-Statistics Division, 1997). Today, poverty and vulnerability affects almost 48 per cent of the total urban population.

In the last 25 years, Quito has experienced a great demographic and spatial expansion since the nineties planning has become increasingly difficult. The intensification of urban environmental conflicts has given rise to many demands for better administrative and planning performance, emanating from the civil society, NGOs and individuals. Associated with these growing demands, efforts by the municipality to increase participation, accountability and democratisation in the planning mechanisms are emerging. One such efforts has been undertaken by the Metropolitan Municipality of Quito under the formulation of strategic plans to be implemented at the 'zone' level within the city. In a decentralised manner, the intention of these plans is to address the growing number and heterogeneity of urban problems, among them the environmental ones. A second event that symbolised a move towards greater attention to environmental matters was the creation of the Environmental Unit.

In this context, in 1996, working as a consultant for the Environmental Unit, we were confronted with several questions concerning how to mitigate the environmental impacts of one of the city's largest socio-environmental conflicts: the contamination of the Machángara river, one of the main water courses running along the city's perimeter. The mitigation of this socio-environmental conflict is quite complex, because it relates to a larger urban problem the city confronts: a highly irregular and unequal territorial distribution. Due to its topography, including mountain ranges, crevasses and valleys, the city's territorial growth has been restricted to a north-south axis. The result of such a spatial distribution has resulted in large segments of poor people living in deficiently equipped neighbourhoods at the outskirts or in environmental and disaster risk areas, such as the margins of the Machángara river.

20.4   Deriving from Experience: The Machángara River Conflict

Lately, the Municipal government has accumulated extensive information and technical knowledge about the sources, types and effects of the Machángara river's contamination (GWE-GTZ and Quito Metropolitan Municipality, 1991). Out of these assessments, it has become evident that the final solution for the river's contamination is to implement a comprehensive sewage treatment and control system, which is projected to be accomplished in a medium to long term planning horizon of 10-20 years. However, despite the municipality's commitment to solve this environmental problem in the longer term, its contamination poses a great threat to the surrounding population in the present.

In the southern end of the city, approximately 100,000 people have direct contact or access to the river, because it runs on the surface for about 8 km of its total

15 km course through the municipality.[2] Paired with such environmentally grave conditions, preliminary research has revealed that the majority of the population living there fall below poverty line measures, and lack access to one or more of the basic needs.[3] In general, sanitation and sewage disposal infrastructure is non-existent or in very precarious condition, and most of the effluent discharges being both organic and industrial, are done directly unto the river course.

Environmental and health conditions in the area have deteriorated progressively. Reports suggest a greater incidence of skin and stomach diseases among inhabitants of the riverine neighbourhoods (GWE-GTZ and Quito Metropolitan Municipality, 1991). Also, especially during the dry season of July-September, the river expels very strong and fetid odours, and large amounts of all sorts of garbage can be found in the river banks.

Affected by a grave deterioration in environmental and social conditions, economic rehabilitation programmes have been deterred for investing in this area. Both public and private investors have abandoned any plans to promote social, economic and environmental recovery. Due to this, land values are comparatively less than in the rest of the metropolitan area, which is the attracting feature for both poor settlers and hazardous industries that evade pollution controls (Carrión, 1987a).

20.5 Local Indicators as a Mechanism to Manage the Machángara Conflict

Keeping in mind that the final solution to the river's contamination will only be achieved after a long period of time and involving very large capital investment, the Environmental Unit decided to propose local indicators as a strategy to mitigate some of its effects in a shorter period. It was also concluded that participatory indicators could at the same time involve stakeholders participation and promote a greater environmental awareness.

The selection of this strategy had three main intentions. On one hand, this strategy was chosen as feasible because it could contribute to close the information gap between the Municipality and the stakeholders. As stated previously, local officials have an extensive technical knowledge about the Machángara's problems and solutions. However, this information has not been effectively communicated to the population, who instead of asking for integrated solutions want to have the river's course 'sealed with concrete' (Personal communication, 1996). Also, although the environmental problem is acknowledged and experienced by everyone in the vicinity, it has not generated any qualitative changes in the population's behaviour. People continue to contribute to greater environmental damage for example by depositing their garbage in the river banks. In addition, during the field recordings many residents were observed collecting water from the river and using it to clean their backyards.

A second reason for selecting indicators was that, through this strategy, stakeholder participation could be greatly enhanced and their own experiential expertise could be recognised. By involving the stakeholders in the process of constructing and implementing local environmental indicators, an effective monitoring system could develop. Once the population embarked on the indicator process, they could provide the municipality with systematic information on the quality of the river. Also, the idea was to develop 'local audits' in which the individuals would be able to monitor the discharges made unto the river in a qualitative manner. Guided by an indicator checklist observable, changes in the river's contamination could be reported.

Finally, this strategy was conceived as a consciousness raising and education starting point, aimed to transform the populations' perception and behaviour towards environmental resources, in this case the river. By incorporating the community in the process, the individuals will acquire a greater knowledge of their environmental surroundings and an empowering process could be generated. Through a translation, confrontation and evaluation process, the indicators generated at the local level would foster a two way communication process. In turn this could lead to the formulation of accountable and legitimate urban planning decisions and policies. Crucial to this strategy was to endorse the local expertise as well as the municipal-technical expertise, and to develop a participatory mechanism reconciling scientific rigorousness in the form of environmental standards and measurements with resonance, especially local acceptance (de Wel, 1995).

20.6 Conclusions

Due to their intricacy, urban-environmental problems, such as the Machángara river basin contamination, need to be addressed from a multidisciplinary and synergistic perspective. Previous top-down designs have been based on unilateral logics which have failed to incorporate local participation from NGOs, and civil society in general. Without considering participation as a key asset to generate effective planning and monitoring mechanisms, many efforts have failed. Nevertheless, the drive to legitimise planning designs by consultation or partnership processes with the population can actually lead to better and more everlasting results.

As the preliminary observations from the Machángara river indicator project suggest, the population has become more aware about the sources, effects and impacts of the river's contamination. As the first sessions with the stakeholders suggested, an attitudinal change had been initiated, especially in the way in which they have addressed the conflict. It was no longer only the responsibility of the municipality but also theirs.

In the case of the Machángara, it was evident that a weak point was the inability to effectively transfer environmental information to the population. Since the project is still in a preliminary phase, conclusive results are not available. However, from the start it was observed that involving the community in the construction of indicators was a very effective mechanism. For the Municipality, it helped to convey and communicate information in a more effective manner than have other environmental education programmes implemented in the area. Also, for the stakeholders, it was a very positive experience because they were able to express their opinions and perceptions about the problem. Overall, the process is fostering a two-way communication and negotiation process, which in the future will perhaps contribute to the formulation of comprehensive and legitimate planning designs.

Notes

1   Sigrid Vásconez D. is Senior Consultant for CHRYSALIDA Consultoría Integrada, Quito, Ecuador. She has experience in urban environmental management and environmental education, with local governments and local NGOs. She completed the Master's programme 'Politics of Alternative Development Strategies' at the Institute of Social Studies, The Hague-Netherlands. Her main research is focus on environmental issues and democracy in Latin American cities.

2   After reaching the colonial center of the city (km 11) the river flows along a steep crevass that impedes direct contact.

3   Field observations and preliminary interviews with residents of five of the riverine neighbourhoods were conducted during a month prior the proposal presentation. Field research was conducted in an attempt to obtain a better understanding of the social and environmental problem.

References

Burgwal, G. (1993) *Caciquismo, Paralelismo and Clientelismo: The history of a Quito squatter settlement, Urban Research Working Papers: 23*, Institute of Cultural Anthropology/Sociology of Development, Vrije Universiteit, Amsterdam.

Carrión, F. (1987a) Balance General de la investigación urbana en el Ecuador in F. Carrión (ed.) *El Proceso Urbano en el Ecuador*, ILDIS, Quito, pp. 11-36.

Carrión, D. (1987b) La renta del suelo y segregación urbana en Quito in F. Carrión (ed.) *El Proceso Urbano en el Ecuador*, ILDIS, Quito, pp. 81-122.

Franco, R. (1995) *Métodos para la medición de la pobreza*, División de Políticas Sociales-CEPAL, Santiago, Chile.

Gilber, R., D. Stevenson, H. Girardet and R. Stren (1996) *Making Cities Work: The role of Local Authorities in the Urban Environment*, EarthScan Publications, London.

GWE-GTZ and Quito Metropolitan Municipality (1991) *Resumen Ejecutivo: Plan de Saneamiento Ambiental para el río Machángara y Monjas*, Municipio Metropolitano de Quito, Quito.

Hammond, A., A. Adriaanse, E. Rodenburg, D. Bryant and R. Woodward (1995) *Environmental Indicators: A systematic approach to measuring and reporting on environmental policy performance in the context of sustainable development*, World Resources Institute, Washington D.C.

Hoshino, C. (1994) Land development processes and decentralization in Latin American large cities and metropolitan areas: issues, trends and prospects, *Regional Development Dialogue*, Vol. 15, no. 2, Autumn, pp. 29-60.

Larrea, C., and M. García (1995) La Situación Habitacional en el area urbana en el Ecuador: 1962-1990, Paper presented in the Workshop *Análisis de Indicadores Urbanos, de Vivienda y Medio Ambiente en el Ecuador*, organised by the Urban Development and Housing Ministry, RHUDO/SA, USAID, Interamerican Planning Association, Catholic University, CIUDAD, Quito.

UNFPA-Statistics Division (1997) *Social Indicators*, in:
http://www.un.org/depts/unsd/mnsds/mnsds.htm.

Wel, B. de (1995) *Indicadores Locales de Sustentabilidad*, Instituto de Ecología Política, Santiago, Chile.

Chapter 21

The City Bubble:
A Framework for the Integration of Environment, Economy and Spatial Planning

V.M. Sol[1], J. de Boer[2], F.H. Oosterhuis[3], J.F. Feenstra[4] and H. Verbruggen[5]

21.1   Introduction

Up till the end of the 1980s the environmental problems associated with cities were seen as relatively small local problems with short term effects. Urban planners approached these problems with effect-oriented measures. Polluted water was discharged into major waters and waste was transported to rural areas, environmental zoning prohibited the influence of industrial sites on the living areas.

Nowadays, urban environmental problems are acknowledged to be much more complex, diffuse sources such as traffic play an important role and have both a local and global effect on the short and long term (Stanners and Bourdeau, 1995). These problems can no longer be solved by enlarging the distance between source and potential effect in time and space. Especially in The Netherlands, space is scarce. This space has to be allocated for residential areas (16 million people and increasing), infrastructure (increasing mobility), economic developments (such as the important developments of Schiphol airport and Rotterdam-Rijnmond harbour), and the development and conservation of nature (Government Planning Agency (RPD), 1993). All these developments claim more space.

The Dutch regional and urban planning aims at concentration of activities (housing, employment, recreation) in the cities in order to be able to protect existing and develop new 'green' areas with a high environmental quality. The objective is to prevent The Netherlands from becoming covered by a so called 'grey blanket', meaning that the environmental quality is equally divided over the country with no very good or very bad areas. The consequences of this policy are different at the regional and the local level. At the regional level, the concentration of activities leads to a relative decrease in environmental pollution as a result of greater efficiency (e.g. by application of the total energy principle and scale economies, and reduced transport). By contrast, at the local level the absolute pollution may increase due to this concentration of activities, often beyond the generic standards (Ministry of

Environment, 1993). This effect obstructs the urban environmental protection policy, that has to comply with the Dutch environmental policy. This national policy aims at legal standards that can be applied nation-wide. A number of large cities in The Netherlands are already having problems in meeting these standards, and concentration of activities increases these problems. Moreover, finances available for pollution abatement measures are limited. The current regulatory approach however, offers no possibilities to balance different environmental problems, such as noise annoyance and air pollution, against each other and against different economic interests (housing, industry, etc.). Therefore, Dutch urban planners have increasingly been arguing that these command-and-control regulatory approaches are not effective and that another, economic approach will give better results, especially for the quality of the environment. They claim that economic incentives are better to achieve environmental quality goals. Also in the national policy there have been signs that another phase of the environmental policy may be at hand. This includes a shifting of regulation from the national level to self-regulation within regional/local administrations, with more responsibility and self determination. An example of this change is a project started by the Environmental Ministry in co-operation with provinces and municipalities, in which the possibilities of applying concepts such as differentiated environmental standards are being studied (Ministry of Environment, 1993).

In this new phase of environmental policy new instruments have to be developed. The Institute for Environmental Studies (IES) contributed to this process with two studies. The first, commissioned by the Environmental Department of the City of Amsterdam, was aimed at the development of a methodology that represents a more economic approach of the 'paradox of the compact city' (Rosdorff et al., 1994). This paradox can be described as follows: generic standards for a local situation obstruct the policy of concentration of activities, and thus obstruct the realisation of decreasing environmental pollution at a regional level. As a result of this study the concept of the 'city-bubble' was presented as a method to be used by local administrators to strike a balance between the development of activities (concentration of functions) and the improvement of the local environment (Sol and Rosdorff, 1994). The 'city-bubble' has been derived from theories about emission trading and the application of an integrated environmental index.

Emission trading is the most important application of the economic instrument marketable permits, which is mainly applied in the USA, primarily involving air pollution (Hahn, 1989). One form of emission trading that is increasingly being used is the (alternative) emission reduction approach that is known popularly as the 'bubble concept' (Tietenberg, 1980). In this case an imaginary bubble is placed over a factory to cover all emissions coming from different sources within the factory. As long as the sum of the emissions is not exceeded the firm can adjust the levels of control applied to different sources. Hahn (1989) concluded that emission trading has afforded many firms flexibility in meeting the emission limits and that this flexibility has resulted in significant aggregate cost savings, worth billions of dollars. The effect on environmental quality has been neutral or slightly positive. It is a useful instrument when the application of generic emission standards can seriously obstruct economic development, and in situations in which maximum ceilings to total pollution are urgently required (OECD, 1994; Anon., 1994).

These experiences were the motivation to combine this economic instrument (emission trading) with an instrument developed by the IES in an earlier stage for application in land-use zoning, the Integrated Environmental Index. The Integrated Environmental Index serves as a representation of the general environmental quality on a specific location (Sol et al., 1995). It reflects in one figure all the possible adverse consequences caused by simultaneous exposure to several environmental pollutants. This chapter will not go into detail on this index. More information can be found elsewhere (De Boer et al., 1991; Sol et al., 1995). By means of the IEI several environmental pollutants (noise, odour, risk, air pollution) can be incorporated in the bubble. The combination of both methods was given the name 'city-bubble' concept. It symbolises that within the city of Amsterdam certain trade-offs are allowed as long as the total pollution decreases over time.

The imaginary city-bubble that is placed over the city refers to the reference level of local environmental quality. Decisions in city planning result in changes compared to this level that can be considered as loss or profit for the inhabitants. Profits occur when houses in a compact city get a better environmental quality than the reference level. Houses with a worse environmental quality experience loss and should be compensated. This will be discussed in greater detail below.

Revealing the profits and losses of certain decisions in city planning makes it possible to pursue a policy that offers the highest profit against the lowest costs. The policy aim is to lower the total cumulated environmental pollution (bubble) in time.

The results of this first study were gladly accepted. The philosophy has already been adopted by the Amsterdam Environmental Department and the Department of Urban Planning in a number of policy documents and brought into the public in several Dutch planning magazines by these departments, but it was also recognised that the method needed more scientific background and support. Also, it is generally agreed that next to environmental pollutants such as noise and air pollution, other issues (crime, public transport, clean appearance, etc.) are important for urban quality of life. Therefore, a follow-up study was performed with two objectives (De Boer et al., 1996). The first is to develop the underlying theoretical concept of the method. This also means that the possibilities for incorporating aspects of liveability and sustainable development should be included, as these are two important policy goals of the city of Amsterdam. As a second objective the concrete possibilities of the method as an instrument of decision making had to be examined. These two elements will be discussed below.

21.2 Theoretical Background

The base of the city-bubble concept is the economic welfare theory. According to this theory individuals are continuously making trade-offs to maximise their welfare. As a result, their welfare is the sum of both positive and negative components. Examples of negative components are the costs involved for obtaining goods or the adverse effects due to activities of others, e.g. noise of car traffic. Each individual balances these components against each other. This way, one citizen chooses a house in a very quiet residential area with a low level of facilities, another may prefer to live in a very noisy area with a high level of facilities. In view of their preferences both achieve their maximum welfare under the given circumstances. It follows that welfare does not only include income, but also environmental quality, the quality of the house, and other aspects of quality of life. The welfare of a city addresses the collective welfare of all individuals in a society. Government is assumed to take into account the interest of all citizens when making decisions. It should express both the desires of citizens and its own objectives for assuring a liveable environment. Therefore, public goods are also a component of the collective welfare. Public goods have the property that their use by one person does not exclude use by others. Examples of goods with these properties are streetlights, parks, clean air, etc. (Daly and Cobb, 1989). Although these public goods are beneficial and wanted by individuals, they are often a free accessible resource and nobodies property. This results in environmental problems, such as air pollution. To avoid abuse of resources government has set up rules, such as restrictions of emissions, but the resources itself are not considered. The resources are freely accessible for many users and the quality is not defined. Therefore, the dimensions, composition and availability of these resources should be fixed. This is the theoretical rationale for the bubble concept. Using the city-bubble concept the environmental quality is described enabling management of an otherwise free available resource.

The welfare theory also implies the compensation principle, that plays an important role in the city-bubble. It declares that a state Y is socially preferable to an existing state X if those who gain from the move to Y can compensate those who lose and still have some gains left over (Dasgupta and Pearce, 1972). In this case: The new city-bubble is preferable to the existing one if the new city-bubble is lower, i.e. the total pollution in the new situation is lower, although some citizens may experience a loss that has to be compensated. Compensation is a principle that is becoming more and more important in Dutch policy plans on environment, traffic and planning. In the country planning document on the Dutch green areas (Ministries of Agriculture and Environment, 1992) compensation in kind (development of new ecosystems, marsh areas etc.) and financial compensation are discussed. Several forms of compensation are also possible in the city-bubble. The first, compensation in kind, refers primarily to compensation between different forms of environmental pollution. However, assuming the welfare theory, a bad environmental quality can be compensated to a certain extent by other aspects of quality of life. It has been noted in literature that the attractiveness of a neighbourhood for residents can constitute a certain compensation for the appearing environmental pollution (e.g. Appleyard, 1981).

21.3   A Framework for Collective Welfare

In order to be able to balance different components of welfare against one another these components need to be specified. A set of indicators was developed, following Button and Pierce (1989), who stated that the urban welfare depends on the quality of life and on the economic base, a proxy for which might be urban real income.

Therefore, at the highest level, collective welfare consists of economy and liveability. These two categories include a number of components that can systematically be itemised in order to build a framework. In the category of economy a distinction can be made between functional and physical qualities. The first concerns the value companies add to the municipal economy in terms of income and jobs. The physical qualities concern the extent to which the economic activities take up scarce ecological resources, space, and transport possibilities. These qualities are related to sustainability.

Liveability points to the physical and social-spatial conditions that should be present at a certain place enabling human beings to live there healthy and well. These conditions can be described as spatial quality, environmental quality and quality of houses. Within spatial quality a distinction can be made between the use value of physical elements (the functional quality, e.g. the number of facilities in an area,) and the perception of the physical elements of an area (physical quality). Figure 21.1 gives a diagram of the identified components of collective welfare.

		Functional qualities economy
	Economy	
		Physical qualities economy
Collective welfare		
		Functional spatial quality
		Physical spatial quality
	Liveability	
		Environmental quality
		Quality of houses

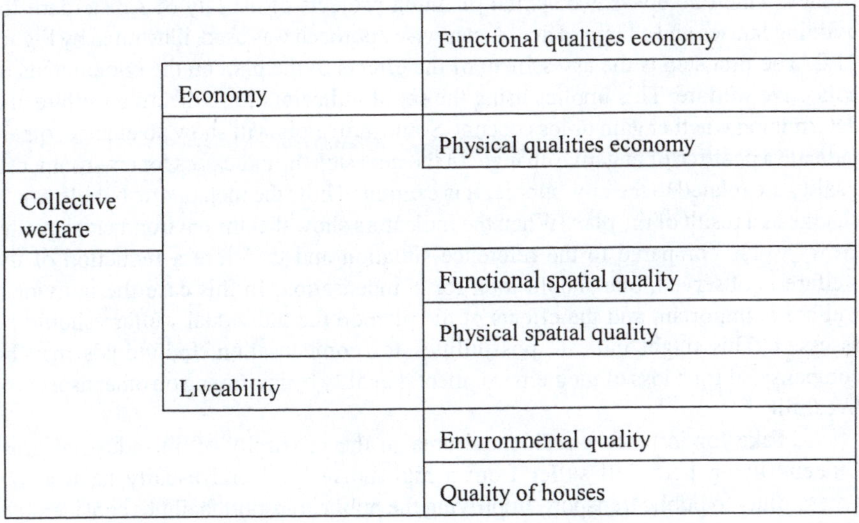

Figure 21.1   Components of collective welfare

The identified components can be split into different aspects, for which indicators can be formulated. As it was not an objective of the study to compose a complete set of indicators, we will not pursue this issue here any further. The following examples illustrate some possible indicators. The amount of m^2 green area and the number of facilities are possible indicators for the functional spatial quality, the number of houses that are built in a eco-efficient way and the m^3 living space per person may be indicators for quality of houses.

21.4   Components of Individual Welfare

A similar procedure can be followed to identify the components of individual welfare. For an individual the socio-economic position (with financial and non-financial components) replaces economy, the other components are equal to those of the collective welfare.

Breaking down the components into aspects and indicators, the aspects of the individual welfare are equal to those of collective welfare. However, the indicators are different, because another level of scale is used for the individual person. For instance, in the case of facilities it is of importance for a citizen whether the facility is close to his/her residence.

21.5   Use of the Framework for Decisions

The second objective of the study was to examine the concrete possibilities of the city-bubble as an instrument for urban decision-making. This was investigated in two case-studies, which involved two spatial planning projects of the city of Amsterdam for building houses and a large office. A stepwise approach was used, illustrated by Figure 21.2. The first step is the assessment of the effects of the plan on the components of collective welfare. This implies using the set of indicators for collective welfare and determining whether gain or loss occurs. Some indicators will show no change, others indicate a positive or negative change. In the next step the indicators for environmental quality are related to the city-bubble. It is examined how the metaphoric bubble would change as a result of the plan. When the indicators show that the environmental quality grows worse compared to the reference situation and therefore a reduction of the welfare is observed, one should look for compensation. In this case the individual welfare is important and the effects of the plan on the individual welfare should be assessed. This might indicate possibilities for compensation. Individuals may be compensated for a loss of their environmental quality by improving on other aspects of liveability.

Take for instance a residential area in the proximity of an industrial area. Citizens living here will suffer from a high noise level and usually have a low accessibility to public transport. Improving the public transport facilities considerately will increase their welfare. This may also have a positive effect on the use of private cars and therefore on the noise and air pollution caused by these cars and improve the environmental quality. In the case that the environmental quality (bubble) of the city does not improve it should still be compensated by measures in other areas of the city. This will have consequences for the costs of implementing the plan. Therefore, also an analysis was made of the available instruments that can be brought into action in a policy based on the city-bubble concept. These instruments should have a number of properties to manage public goods, e.g. be able to generate financial means that can compensate individuals for the change in availability of a public good. Market based (economic) instruments may provide these possibilities. Examples include taxes (e.g. for parking, houses, waste processing), subsidies (stimulation of environmental conscious behaviour or compensating victims), fees, charges, marketable permits and covenants (James et al., 1978).

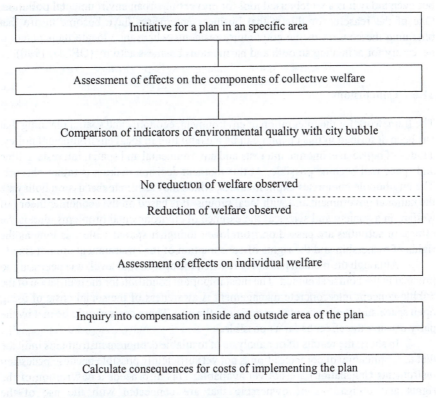

Figure 21.2   Use of framework for decisions

The question of who has to pay has different answers, depending on the situation. As a first example, consider advancing residential buildings in an area where a company has been present for a long time and has been keeping to the rules of its licence. It seems only reasonable to charge costs of compensating or mitigating measures to the new residents, although the economic feasibility of the project should be certain. The property developer and intended residents should be willing to pay (i.e. 'the user pays principle'). In the case the polluter has to pay, but cannot comply to standards, for technical or financial reasons, the company might finance compensating measures. The present Dutch environmental legislation does not offer many possibilities to include these forms in the licence regulations. Another possibility might be a covenant in which local government, business and residents negotiate on agreements in order to prevent pollution. This instrument has not been used very often in the local Dutch environmental management, contrary to Japan, where 'anti-pollution agreements' are signed on a regular base. Anti-pollution agreements are contracts between local authorities, inhabitants and a business sector to reduce or prevent pollution caused by local business activities. The first (starting in 1952) were used to fill the gaps created by the absence of adequate pollution related laws and regulations, but even today it is a widely used tool for preventing urban environmental pollution. One of the reasons for this is that business enterprises have become aware that obtaining the consent of the community residents where they locate their plants is necessary for achieving smooth and harmonious business activity (OECD, 1990).

21.6   Conclusions

The framework for the integration of environment, economy and spatial planning that has been described in this chapter can be a useful tool in local environmental policy. Trade-off between economic interests and environmental and spatial interests in time and space will become possible. At this moment decision making is rather obscure. The city-bubble comprises explicit values, which may facilitate discussion both with the national government and with local groups. It is based on the economic theory of welfare in a society and seizes at the essence of environmental problems, that is the effects of activities are passed on from lower to higher spatial scales, as long as the rights of ownership and the rights of consumption of the environment are not fixed.

Although the method has never reached the operational level, it is necessary to proceed in the course sketched. The most important condition for the translation of the bubble-concept into concrete management is awareness of the social costs of using open space and environmental resources in the city. These costs should be met by the party causing the effect as far as possible.

In short, the results of our analysis of available economic instruments indicate that the application of compensation is not yet sufficiently possible because necessary instruments (legislation) are not yet available. This also needs a description of the rights and obligations of ownership that are connected with the use of the environmental resources. At this moment such a description can not offer the same legal security as standards. It should be prevented that agreements between local government, business enterprises and residents can be reversed at a later moment. A regulation for disputes is therefore necessary.

Notes

1   Dr. Vera Sol is an environmental chemist at the Institute for Environmental Studies, Vrije Universiteit Amsterdam. Her main activities relate to studies supporting environmental policy, with special attention to integration of environmental impacts in urban areas.

2   Dr. Joop de Boer is research co-ordinator at the Institute for Environmental Studies, Vrije Universiteit Amsterdam. His current research includes community response to environmental stressors and threats, especially in the field of risk perception and risk communication.

3   Dr. Frans Oosterhuis is an environmental economist at the Institute for Environmental Studies, Vrije Universiteit Amsterdam. He is presently mainly involved in research

projects on energy policies, economic instruments in environmental policy and environment-trade relationships.

4   Dr. Jan Feenstra is deputy head of the National Sciences Department of the Institute for Environmental Studies, Vrije Universiteit Amsterdam. He is mainly involved in Climate Change studies and studies founding international, national, provincial and urban environmental policy.

5   Prof. Harmen Verbruggen is head of the Department of Economics, Technology and Social Sciences of the Institute for Environmental Studies, Vrije Universiteit Amsterdam. His major field of specialisation is the interplay between development economics, international economic relations and environmental studies.

References

Anonymous (1994) Emission cuts, *New Scientist*, 141, March 12, p. 11.

Appleyard, D. (1981) *Livable streets*, University of California Press, Berkeley.

Boer, J. de, H. Aiking, E. Lammers, V. Sol and J.F. Feenstra (1991) 'Contours of an integrated environmental index for application in land-use zoning' in O. Kuik and H. Verbruggen (eds) *In search of indicators of sustainable development*, pp. 107-20, Kluwer Academic Publishers, Dordrecht.

Boer, J. de, V.M. Sol, F.H. Oosterhuis, J.F. Feenstra and H. Verbruggen (1996) *De stadsstolpmethode – een afwegingskader voor de integratie van milieu, economie en ruimtelijke ordening bij stedelijke ontwikkeling*, Milieudienst Amsterdam, Amsterdam.

Button, K.J. and D.W. Pierce (1989) Improving the urban environment: how to adjust national and local government policy for sustainable urban growth, *Progress in Planning*, 32, pp. 135-84.

Dasgupta, A.K., and D.W. Pearce (1972) *Cost-Benefit Analysis: Theory and practice*, MacMillan, London.

Daly, H.E., and J.B. Cobb (1989) *For the common good. Redirecting the economy toward community, the environment and a sustainable future*, Beacon Press, Boston.

Government Planning Agency (RPD) (1993) *Ruimtelijke verkenningen 1993* (Spatial explorations) Ministry of Housing, Physical Planning and Environment, The Hague.

Hahn, R.W. (1989) Economic prescriptions for environmental problems: how the patient followed the doctor's orders, *Journal of Economic Perspectives*, 3, pp. 95-114.

James, D.E., H.M.A. Jansen, and J.B. Opschoor (1978) *Economic approaches to environmental problems. Techniques and results of empirical analysis*, Elsevier, Amsterdam.

Ministry of Agriculture, Nature Management and Fisheries, and Ministry of Housing, Physical Planning and Environment (1992) *Structuurschema Groene Ruimte* (Structural Scheme for the Green Open Space) Ministry of Housing, Physical Planning and Environment, The Hague.

Ministry of Housing, Physical Planning and Environment (1993) *The Netherlands National Environmental Policy Plan 2. The environment: today's touchstone*, Ministry of Housing, Physical Planning and Environment, The Hague.

OECD (1990) *Environmental Policies for Cities in the 1990s*, Organisation for Economic Co-operation and Development, Paris.

OECD (1994) *Managing the environment. The role of economic instruments*, Organisation for Economic Co-operation and Development, Paris.

Rosdorff, S., L.K. Slager, V.M. Sol and K.F. van der Woerd (1994) *De stadsstolp: meer ruimte voor milieu èn economie* (The city dome: more space for environment and economics) Report R-94/3, Institute for Environmental Studies, Free University, Amsterdam.

Sol, V.M., and S. Rosdorff (1994) *A framework to trade-off economic and environmental interests for the city of Amsterdam*, Paper presented at the First International Expert Seminar on Advanced Environmental Management Tools and Environmental Budgeting at a local level, 14-16 March, Freiburg.

Sol, V.M., P.E.M. Lammers, H. Aiking, J. de Boer and J.F. Feenstra (1995) An integrated environmental index for application in land-use zoning, *Environmental Management*, 19, pp. 457-67.

Stanners, D., and P. Bourdeau (eds) (1995) *Europe's Environment: The Dobris Assessment*, European Environmental Agency, Copenhagen.

Tietenberg, T.H. (1980) Transferable discharge permits and the control of stationary source air pollution: a survey and synthesis, *Land Economics*, 56, pp. 391-416.

Chapter 22

An Endeavour at Integration in Environmental Analysis and Planning

M. Partidário[1] and H. Voogd[2]

22.1   Introduction

In all sectors of our society a growing complexity can be witnessed. We are using more and more complex machines for 'making our life easier'. We don't travel any more by horse but by car, and not by stage-coach but by train or aeroplane. We don't need to collect and burn wood or coal for heating meals any more, but are now able to do the same job in seconds by means of a microwave. We don't use typewriters any longer, but we use computers with 'word processing programs'. And these computer programs become more complex over the years, integrating more and more functions. In order to make sure that ordinary people can handle them, computer programs, cars and other complex machines are being designed in such a way that their operation is very simple. Much attention is devoted to the prevention of undesirable situations, i.e. a 'perfect' machine or computer program should be *fool proof*. The ultimate aim is that full operation is possible by just pressing a few buttons and that nothing can go wrong if the wrong buttons are being pressed. Evidently, the growing complexity of society seems to go in hand with a growing need for *simplicity* and *surveyability*.

The desire for 'simplicity' can also be found in public planning and policy-making. The words 'selective approach' or 'selectivity' are often used to express the device 'rather partially good than entirely wrong'. Indeed, there is nothing to be said against focusing on 'main points', but who determines what a 'main point' will be? And how 'fool proof' is a selective approach? Can we still prevent disasters if a selective approach goes wrong? Also a desire for 'surveyability' can be noticed in planning and policy-making, for instance by the use of phrases like 'integral' or 'comprehensive' or alike. Especially in environmental planning there is a growing concern that environmental problems cannot be treated in isolation, but that interrelationships between environmental problems as well as with social and economic problems should be included (e.g. see also Miller and de Roo, 1997; Voogd, 1994). Evidently, this calls for an 'integrated approach', but what does it actually mean?

The purpose of this chapter is to elaborate on this quest for integration. In a sense this means a return of intellectual concepts such as 'holistic thinking' and 'comprehensive planning'. This will be discussed in the next section. In section 22.3

attention is paid to some procedural-organisational forms of integration. Section 22.4 is devoted to a innovative research project conducted at the New University of Lisbon (Partidário, 1996), which attempts to develop a comprehensive methodology for the effective consideration and integration of natural and physical resource policies with land-use planning policy and strategies. In section 22.5 some remarks are made about the role of Strategic Environmental Assessment as a tool for integration. This chapter finishes with some concluding remarks.

22.2   The Concept of 'Integration'

In philosophy comprehensiveness has received a lot of attention under the heading of *holism* (e.g. see Gauker, 1988; Healey, 1991). The overwhelming complexity of man's place in the world, the threat of global warming, depletion of the ozone layer, pollution of water, air and soil, developmental pressures at the regional and local levels are all examples of increasingly complex systems. In all of these cases, man's involvement within the system plays a large role. Advocates of holistic approaches such as Senge (1990) and Marietta (1995) argue that only when we have a better grasp of the systems as a whole can we begin to understand how we can organise possible solutions for environmental problems.

A most flexible description of holism is that it is the denial of particularism (Healey, 1991) or individualism (Kincaid, 1993). Holism can be defined as a philosophical doctrine which implies that there are properties of a system which can be better understood by investigating them independently of the properties of its parts (Liu, 1970). Clearly, individualists disagree by imposing the methodological norm that purely individualist theories suffice to explain fully, or – even more strict – full explanation requires reference solely to individuals (see Kincaid, 1993).

There have also been many pleas in the past for comprehensiveness in planning and planning analysis. According to John Friedman (1987, p. 124) different words have been used in the past by planners to emphasise this: Karl Mannheim (1951) speaks of *interdependent thinking* grounded in specific situation, Rexford Tugwell (1935) uses the metaphor of a *collective mind* capable of overcoming the partial and fragmented knowledge of disciplinary specialisation's and also Lewis Mumford (1938) stressed the need for *simultaneous thinking*. Evidently, this shows that the quest for more integration in analysis and planning is not a recent phenomenon.

The comprehensiveness paradigm has also been seriously criticised in the past, for instance by Braybrooke and Lindblom (1963), Friedmann (1971) and van Gunsteren (1976). The root of the matter according to these authors is that full comprehensiveness can never be realised, that it is a very inefficient approach because of the 'ad hoc' nature of actual policy-making, and that more specific and flexible approaches are therefore more appropriate. However, not everybody was convinced by these arguments. For instance, Perloff (1980) again called for an integrated approach of environmental, physical, economic and social-political analysis, which was according to Friedmann (1987) '*a game plan for four-dimensional chess*'. Obviously, an integrated approach is very difficult to pursue, but should we therefore restrict ourselves to simple approaches that ignore the existing complexity? Can we afford to be 'narrow-minded' or should we focus more on making integrated approaches more '*fool proof*', analogously to the design of complex machines?

We feel that it is essential, given the complexity of the environmental system and despite the many practical difficulties, at least to study the possibility of arriving at more integrated analytical and planning approaches. Evidently, the complexity of the environmental system should be reflected, in one way or the other, in the *substantive* planning approach, i.e. focusing on the content of the problem. In section 4 an illustration is given of a research project that is aiming to develop a such methodology for integrated environmental and land use planning analysis (cf. Partidário, 1996). Besides substantive issues, also *procedural* planning issues need to be considered. Different kinds of 'environments' can be distinguished, which are often dealt with by different actors with different cultural backgrounds, different interests and different objectives. Integration in this respect means a collaboration or organisational amalgamation of actors or groups of actors, harmonising rules and procedures, bridging gaps between cultures, e.g. the 'engineering culture' and the 'environmentalist culture' or the 'economist culture' (cf. Voogd, 1995). This is discussed in more detail in the next section.

22.3 Integration as a Procedural Planning Problem

Public planning, and as such also environmental and land-use planning, has a bureaucratic practice as well as a political practice. Bureaucratic practice is articulated through the institutional structures, whereas political practice is exercised by the politically active community (cf. Friedmann, 1987). In order to be acceptable for these practices, planning approaches should be adapted to the characteristics of both kinds of practices. And this will be a difficult task, due to the fundamental differences both between the bureaucratic practice and the political practice and within both practices. Evidently, integrated planning implies an attempt to overcome or to deal with these differences.

In most countries bureaucratic practice has caused rather fragmented planning and policy-making structures, often coinciding with sharply defined departmental structures. As a consequence different planning laws and different planning procedures may exist, dealing with different policy areas – policy sectors – for which different political executives are responsible. Hence, it is not unusual that *sectoral conflicts* exist, for instance between environmental policy and economic policy. Evidently, an integrated approach should be able to deal with these sectoral conflicts, that may have a political, organisational, social or even a simple 'individual human nature' background.

In Table 22.1 four different models are presented for arriving at a better procedural integration of sectoral planning, especially planning with spatial implications, such as environmental planning, land-use planning, traffic and transportation planning, water management, and so forth.

Table 22.1   Procedural integration models

	Simultaneous but Separate Sectoral Policy-Making	Simultaneous and Integrated Policy-Making
Sectoral plans	*Autonomy Model*	*Political Visibility Model*
Comprehensive plan	*Deliberation Model*	*Back to the Future Model*

The first model will be called the *Autonomy Model*. This is the most modest form of procedural integration, in that it is aimed at a simultaneous preparation of sectoral plans in order to tune in to each other's proposed policies. This model enables a standardisation of levels of abstraction between the various plans and hence a better insight into the various linkages and possible consequences. It also enables simultaneous public discussions. However, there will be no guarantee that sectoral policies are really being adjusted to each other since there still are separate sectoral plans. Each plan will be defended in the 'political arena' by different political executives, who are eventually only responsible for their own portfolios.

The second integration model is called the *Deliberation Model*. This goes one step further then the Autonomy Model, because now it is aimed at the production of one single comprehensive plan for 'society and environment', which combines all relevant policy areas. This comprehensive plan will serve as an 'umbrella plan', as a point of departure for more operational plans. The advantage of this model is that now an explicit and general framework is available, that is not easy to neglect by sectoral policy-makers. An obvious disadvantage will be that the preparation of such a comprehensive plan, given its far-reaching implications for other policies, will be rather time consuming due to sectoral negotiations, both in bureaucratic as well as the political arenas.

Another, more rigorous way for improving the integration of sectoral policies is to adjust the bureaucratic organisation. Instead of maintaining current 'vertical' sectoral departmental divisions, a new organisational structure is developed with better 'horizontal' relationships, e.g. through project organisations or matrix organisations. Obviously, this has implications for the way governments will act. Traditionally, a sector department will have its own political executive in government (e.g. Minister). In a more integrated civil service, this may no longer be the case. In the *Political Visibility Model* this is 'repaired' by ensuring that simultaneous and integrated plan preparations and policymaking still will result in different sectoral plans serving the political portfolios.

The most courageous integration model is no doubt a model that will reduce the complexity of the planning system. This means in a sense a return to the size and structure of government in former days. The essences of land-use policy, environmental policy, transportation policy and water management are being amalgamated in this *Back to the Future Model* into one single planning structure. Obviously, civil servants will not be in favour of this model, but, for instance, tax payers may have another view.

The procedural integration models shown in Table 22.1 can be elaborated and refined in many ways. Other distinctions can be added, for instance about the structuring of *permit systems*, or about the way *monitoring and implementation* is organised. Attention can also be paid to *multi-level integration*, i.e. the integration of policies expressed at a different geographical scale.

The justification of procedural integration is based not only on the tax payer's relief. From an environmental point of view, the most important justification should be the guarantee that environmental considerations and concerns will become major elements in the policy-making and decision-making process. This necessitates also the development of good substantive integration models, such as outlined in the next section, and well-defined Strategic Environmental Assessments (SEA).

22.4 Integration as a Substantive Planning Problem: Towards a New Methodology

Limited natural and physical resources are most likely to influence land-use planning policies, strategies and criteria, if development is to be enhanced in a future of greater scarcity of natural resources. Land-use planning will need to respect thresholds associated to resources sustainability. The stress imposed on resources by land-use development activities relate not only to the rate at which physical and natural resources are consumed, but also to the degradation of resources and land-use quality.

As such, a good articulation and integration of physical and natural resources management policies with land-use planning policies is needed, involving: a systematic analysis of existing land-use planning, and resource management, policies and legislation; the interpretation of resources – land uses and planning actions relationships; the analysis of its relevance to different land-use planning scales; and the formulation of guidelines, to assist the effective consideration and integration of natural and physical resource policies with land-use planning policy and strategies.

A research project at the New University of Lisbon, Portugal (Partidário, 1996a) is developing a methodology, based on the above rationale, that adopts 8 different categories of natural and physical resources (air, water, land/soil, ecological systems, mineral resources, cultural values, landscape and energy). These are used to explore the relationship of those resources with a formal set of categories of land-uses and planning actions, towards the formulation of sustainable management guidelines.

The exploration of the issue follows a five-step methodology, which is summarised below.

First Step – Identification of Relationships

The purpose is to identify significant relationships of dependency between land uses and actions, and resources. The method was based essentially on expert-judgement but also included identification of existing legal requirements. Relationships were registered in a matrix.

Second Step – Analysis of Sensitivity

It refers to the identification of cause-effect relationships between land uses and planning actions and physical and natural resources management, according to three different sustainable management values: use – the interest and utility of the resource for the land use and planning action; quality – this value was interpreted in two ways: (a) the resource quality required to satisfy the needs of a land-use or planning action; and (b) the way categories of land-use and planning actions can negatively affect or positively contribute to enhance resource quality; and availability – the way categories of land-use and planning actions can negatively affect or positively contribute to enhance the resource availability (quantity and renewability).

The assessment of the value of the resources according to these three vectors: use, quality and availability, enabled the analysis of the resources sensitivity. The method involved expert-judgement by means of a three-round Delphi process. Assessed sensitivity was registered in a specific matrix.

Third Step – Sustainability Analysis

Sustainability analysis aims at evaluating whether the land-use/resources relationships identified, and characterised by a sensitivity assessment, can be sustainable, sustainability being understood here as a measure of effective integration. Such integration will reflect the evaluation of land uses and planning activities feasibility as a function of the use, availability and quality of the resources, in the context of the multi-functionality and uniqueness of resources.

The sustainability analysis results from integrating and interpreting the sensitivity values adopted in step two: use, quality and availability values for each land use/resource identified relationship. Consequences over the quality and/or availability of the resources may result in four different outcomes.

1   The use of the resource by any of the land uses or activities raises no problems.
2   The use of the resource by any of the land uses or activities should be incentivated for the opportunities it may create.
3   The use of the resource is important to certain land uses or activities without quality or availability problems; however, such activities may impose a reduction in the resource quality requirements by other land uses and activities, inducing cumulative and synergism effects on the use of the same resource. Specific management requirements are necessary in this regard.
4   The use of the resource by any of the land uses or activities is likely to raise quality or availability problems, therefore management requirements are needed in that regard.

Consequently, the result of the sustainability analysis can be expressed in the following ways.

•   *Very sustainable* – The land use or planning action, whether using the resource, or not, provides for the enhancement of its quality and/or its availability.

- *Sustainable* – The land use or planning action although using the resource, does not have any effect on its quality or its availability.
- *Critically Sustainable* – The land use or planning action, whether using the resource, or not, reduces its quality but enhances its availability or vice-versa.
- *Not sustainable* – The land use or planning action, whether using the resource, or not, reduces the quality and/or the availability of the resource.

Forth Step – Political Context for Resources Management

Taking stock on the analysis of legislation developed in the 1st step of analysis, this step in the methodology intends to interpret the legal opportunities and incentives, as much as the restrictions and barriers to the use of resources by land uses and planning actions.

Likewise, it seeks to interpret the effectiveness of land use and resource management tools relatively to the sustainability objectives defined. This interpretation will enable the identification and justification of existing barriers in land use planning and resource management policies and point towards ways, actions and measures that can contribute to achieving the desired integration.

Fifth Step – Formulation of Guidelines

Based on the analysis previously conducted, the purpose is to elaborate on resource management recommendations, that can be adopted, as guidelines, by planning practitioners, at different planning scales. Such guidelines are designed to contribute to the preparation, development and management of land use plans at the local and regional levels, holistically and across sectors.

The result will certainly reflect the conceptual understanding of the experts involved in the analysis of specific physical and natural resources and land use planning processes in the context of the project. The calibration of this methodology, with the assistance of a larger number of experts, and referring to a wider variety of policy and planning realities, will certainly improve the validity of its results for wider applications.

22.5   Strategic Environmental Assessment and Integration

Another way to consider substantive integration in planning is by means of a Strategic Environmental Assessment (SEA). Some authors argue that SEA is about integration of environmental assessment principles into the decision-making process (e.g. see Sadler, 1994). Others distinguish clearly environmental assessment from integration (Bregha, 1990; Holtz, 1991). Clearly, '*environmental assessment of policy in its widest sense includes both the systematic integration of environmental considerations into the policy formulation process and the assessment of a policy's environmental effects*' (cf. Bregha, 1990, p.2). To better define SEA in terms of integration, the following typology can be made.

Full integration Environmental factors and concerns are an intrinsic element in the formulation of actions amenable to strategic decisions, and assessment activities are tailored in this process, ensuring continuity of assessment.

Environmental shape Environmental factors and concerns are tailored (integrated) in the formulation of actions but no assessment exists to find out about the importance and magnitude of potential positive and negative effects on the environment during implementation.

Concurrent assessment The action is tailored irrespective of environmental related factors and concerns and a separate environmental assessment procedure is developed simultaneously with the development of the policy.

'Staple' integration An environmental assessment is carried out only once the action is defined and both action and EA reports are 'stapled' and submitted for approval.

Only the concept of full integration is understood to refer to SEA. The third concept – concurrent assessment – can be considered as referring to most of the existing practice with the application of SEA. Ideally, however, it should be developed to achieve a more complete integration. Environmental shape is what is often known as environmental planning or policy-making. Whilst it is fundamental in the SEA process, it does not fulfil the systematic requirements as referred in Bregha's quotation cited earlier. Finally, 'staple' integration illustrates exactly what should not happen in SEA but which, regretfully, has often occurred with project EIA's.

Full integration is the most desirable means by which sustainable development can be achieved (see also Thérivel and Partidário, 1996). But SEA has a distinctive, though temporary role, to play in this process. SEA can help to achieve the integration of environmental issues in the development of policies, planning and programme decisions by forcing the introduction of systematic practices in the identification of relevant environmental issues and assessment of environmental impacts in pre, as well as in post-implementation stages. Once sound environmentally integrated approaches are achieved, then SEA has played its role and may no longer be necessary (see Partidário, 1996b).

22.6   Some Concluding Remarks

Since 'everything is related to everything', there is an obvious desire in planning and policy-making to be as comprehensive as possible. Notions like 'integrated analysis' and 'integrated planning' have been embraced by planners ever since the word 'planning' began to be used. But seldom has a justification been given concerning what is meant by the prefix 'integrated'.

This chapter has illustrated that the concept of 'integration' encompasses a variety of meanings. According to the dictionary the meaning of the word 'integral' is 'whole, not fractional'. One 'integrates' if one 'makes or forms into one whole'. Integration is the act of 'integrating'. Or to dress up this definition in our own words: *integration is to develop a bird's eye view of an entire system, without neglecting essential details.* But is this feasible? Unfortunately, philosophers have not found a satisfactory answer to this question, nor can we. Etzioni's (1968) *mixed scanning* approach of policy-making, i.e. consider fundamental issues and then focus on relevant details, may seem appropriate, but it has of course also its critics (e.g. Camhis, 1979).

The main driving force behind any form of integration is the expectation that the outcome is better than without integration. Substantive integration may result in more reliable outcomes, but it can also give incorrect answers, for example due to error accumulation because of system complexity. Procedure integration may result in positive synergy (i.e. $2 + 2 = 5$), but it can also lead to negative synergy ($2 + 2 = 3$), for example due to bureaucratic or political obstructions. Therefore the concept of 'integration' needs to be carefully considered before it is embraced as the only or least solution for our planning problems. Obviously, integration needs to be applied in a 'tailor made' manner and adapted to each particular situation in order to assume validity.

Notes

1   Maria do Rosario Partidário is Professor at the Department of Environmental Sciences and Engineering, Faculty of Sciences and Technology, New University of Lisbon, Portugal.
2   Henk Voogd is Professor at the Department of Planning and Demography, Faculty of Spatial Sciences, University of Groningen, The Netherlands.

References

Braybrooke, D., and C.E. Lindblom (1963) *A Strategy of Decision: Policy evaluation in a social process*, Free Press, New York.
Camhis, M. (1968) *Planning Theory and Philosophy*, Tavistock, London (1979).
Etzioni, M. (1968) *The Active Society*, Free Press, New York.
Friedmann, J. (1971) The Future of Comprehensive Planning: A critique, *Public Administration Review*, 31, May/June, pp. 43-51.
Friedmann, J. (1987) *Planning in the Public Domain: From knowledge to action*, Princeton University Press, Princeton N.J.
Gauker, C. (1993) Holism without meaning: a critical review of Fodor and Lepore's Holism: a Shopper's Guide, *Philosophical Psychology*, vol. 6, no. 4, pp. 441-450.
Gunsteren, H.R. van (1976) *The Quest for Control, A critique of the rational-central-rule approach in public affairs*, Wiley, New York.
Healey, R. (1991) Holism and Nonseparability, *Journal of Philosophy*, vol. 88, pp. 393-421.
Kincaid, H. (1993) The empirical nature of the individualism-holism dispute, *Synthese*, vol. 97, pp. 229-247.
Liu, C. (1996) Holism vs Particularism: a Lesson from Classical and Quantum Physics, *Journal for General Philosophy of Science*, vol. 27, pp. 267-279.

Marietta, D.E. (1995) *People and the planet: Holism and Humanism in Environmental Ethics*, Temple University Press, Philadelphia.

Miller, D., and G. de Roo (eds) (1997) *Urban Environmental Planning: Policies, instruments and methods in an international perspective*, Avebury, Aldershot.

Partidário, M.R. (1996a) A Integraço de Políticas de Gesto de Recursos Físicos e Naturais no Ordenamento do Território, 3rd Interim Report to DGOTDU-JNICT, New University of Lisbon, Lisbon.

Partidário, M.R. (1996b) Strategic Environmental Assessment: Key issues emerging from recent practice, EIA Review, No. 16, pp. 31-55.

Perloff, H.S. (1980) *Planning the Post-Industrial City*, APA Planners Press, Washington D.C.

Senge, P.M. (1990) *The Fifth Discipline: The art and practice of the learning organization*, Doubleday Currency, New York.

Thérivel, R., and M.R. Partidário (1996) *The practice of strategic environmental assessment*, Earthscan, London.

Voogd, H. (ed.) (1994) *Issues in Environmental Planning*, Pion Ltd., London.

Voogd, H. (1995) Provinciale Omgevingsplanning, *Milieu – Tijdschrift voor Milieukunde*, Vol. 10, No. 4, pp. 189-195.

Chapter 23

Households, Sustainability and Environmental Quality

A.J.M. Schoot Uiterkamp, K.J. Noorman and W. Biesiot[1]

23.1　Introduction

The 20th century has shown a sharply increasing number of environmental problems. Most of these are associated with (over)exploitation of natural resources, generation of waste, accelerated extinction of species and degradation of ecological functions. This resulted in growing environmental awareness symbolised in phrases like 'Sustainable Development' (WCED, 1987) and 'Environmental Quality' (EQ).

Social response to environmental problems has largely been restricted to fighting short-term symptoms and has – at least until the eighties – focused largely on the production side of economic activities. Only lately are environmental research and policies aiming more towards the consumption side of economic activities. Through methods like integral chain management and lifecycle analysis, consumer activities can be linked to the complex pattern of inputs and outputs of the economy (agro-industrial production and transport of goods and services) and thus to their associated environmental consequences.

Households are the basic consumption units. In most western-style countries the number of households is growing faster than the population. Since urbanisation is strongly increasing world-wide, households tend to be located in cities. In 1994 a major five-year interdisciplinary research program entitled HOMES[2] (HOusehold Metabolism Effectively Sustainable) was initiated at the University of Groningen and at Twente University. Within HOMES, researchers from both natural sciences and social sciences disciplines are co-operating to investigate changing relations between household activities, sustainability, and environmental quality.

Following same initial remarks about the meaning of sustainability and environmental quality the metabolism metaphor is introduced and the HOMES program is described. Major findings of the diagnostic phase are presented and their relevance for urban environmental quality is discussed (Noorman and Schoot Uiterkamp, 1997).

23.2　Sustainability and Environmental Quality

The World Conservation Strategy (WCS/IUCN, 1980) emphasised ecological constraints on human activities and advocated appropriate environmental protection and preservation to ensure the sustainable (i.e. human!) utilisation of species and ecosystems. In 1987 the report 'Our common future' (WCED, 1987) stated that present patterns of economic growth are not ecologically sustainable. The WCED underlined the importance of sustainable development as a common ground for long-term environmental policy by defining it as: 'development that meets the needs of the present without compromising the ability of future generations to meet their own needs'. In this view, sustainability-related issues are associated with social structures and human activities aimed at fulfilling human needs, and also with protecting the quality of life and its physical and biological environment.

Since 1987 many authors have presented a range of definitions of sustainability (Pearce et al., 1989; Pezzey, 1989; Pugh, 1996). The concept of sustainability seems to lend itself more to easier for scientific and political debate than for interpretation and implementation.

To some, besides environmental problems, sustainability issues comprise a broad range of social issues such as poverty, equity, social security and peace. Others emphasise more restricted views, in particular the long-term ecological viability of economic processes (cf Opschoor, 1992).

The concept of EQ turns out to be as value-laden as is the concept of sustainability. So, again many interpretations prevail. Some of them make a clear distinction between sustainability referring to social systems and environmental quality referring to natural systems. In this view sustainability-based policies can be pursued more or less independent from EQ policies (or vice-versa). Other interpretations are based upon a more hierarchical relation between the two concepts. In anthropocentric views sustainability is regarded to be the dominant factor determining the margins left to the natural environment. The opposite holds for ecocentric views.

The anthropocentric perspective is shown in the approach to natural resources in traditional economics. Here environmental quality is measured only by a stock of goods that yield a flow of services that ultimately contribute to maximising the net benefits of economic activity. In the ecocentric perspective, ethical, aesthetic, cultural, as well as technical arguments are also seen as valid reasons for conserving natural capital and preserving biological diversity. Environmental quality is described by Opschoor and van der Ploeg (1990) as the degree to which the natural environment corresponds to objectives concerning variables such as biodiversity, integrity of ecosystems and concentrations of toxic substances. This description seems to be in accordance with that of van Diepen and Voogd (1994). Operating from a geographic perspective, they describe environmental quality in terms of 'degree of excellence', 'property' or 'attribute' of a geographical space.

In an earlier definition of environmental quality (Lansing and Marans, 1969) satisfaction is emphasised as one of its key criteria: An environment of high quality conveys a sense of well-being and satisfaction to its population through characteristics that may be physical, social or symbolic.

In a recent PhD thesis (van Poll, 1997), residential satisfaction is seen as a measure of perceived quality of the urban residential environment. Van Poll concludes that environmental quality in this case may be best conceived of as a hierarchical, multi-attribute concept. This is essentially an environmental effects-based view. The key criteria observed by him are experienced satisfaction with the residential environment (i.e. personal dwelling and neighbourhood) and perceived annoyance by environmental factors. In contrast The Dutch National Environmental Policy Plan (VROM, 1988-1989) adopts an expert-based and exposure-based view on assessing environmental quality.

In conclusion, sustainability and environmental quality are value-based concepts correlated with very different views on the relationship between humankind and nature and between present and future generations.

23.3 Metabolism, Households and HOMES

The concept of metabolism is rooted in life sciences. It refers to the material cycles and energy flows determining the viable and continuing organism-environment interactions.

Applying metabolism as a metaphor in environmental sciences stresses that human societies are using large amounts of material and energy in ways analogous to organisms. Resulting material flows in human societies should occur in 'closed' cycles just as in natural communities of other organisms.

The metabolism concept can be combined with the concept of sustainability: Organisms are self-sustaining when they are at equilibrium with their environment. In such a case resources are metabolised in an optimal way; waste is converted into resources. In addition to industrial processes, the metabolism concept can also be applied to consumer activities and households as organisational units of consumer activities.

By focusing on household metabolism, the HOMES program addresses the complex environmental problems of a major and important segment of a western society typified by the Dutch society. The term 'household metabolism' refers to both the demand for resources, i.e. the direct flows of resources through households, and the supply of resources, i.e. materials and energy indirectly required to realise these flows (e.g. mining, production of materials, construction of houses and manufacturing goods). Household metabolism depends on a large variety of factors (e.g. spatial, technical, economic, behavioural and administrative). Designing effective and socially acceptable policy instruments for reducing household metabolic rates and their negative environmental impacts, requires a thorough understanding of the determinants of household consumption and their mutual relationships. It also requires detailed information on possible differences in the 'lifestyles' and resulting consumption patterns of different population-groups.

Figure 23.1 gives an overview of the metabolic flows related to consumer patterns in households. Natural resources are extracted from the physical environment in order to produce goods and services to meet the material needs, wants and non-material aspirations of the population. Currently, only a small fraction of the total amount of the natural resources that have been extracted is recycled. Recycling takes place at both the level of production and that of final demand consumption by households. Reducing the throughput of flows of matter and energy, by closing relevant physical cycles is regarded as a necessary step towards a more sustainable future.

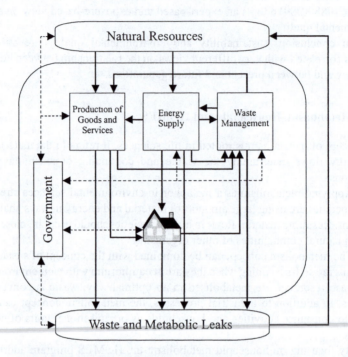

Figure 23.1   The metabolism concept applied to households
Source: Noorman et al. (1997)

Emphasis on households in The Netherlands is justified, as it is a model western society: densely populated, strongly urbanised and with intensive industrial, agricultural and transportation activities and corresponding strong environmental impacts. Along with the number of households, the number of dwellings has increased steadily in the Netherlands. These dwellings have been linked to a steadily rising number of goods and services provided by physical networks, such as drinking water pipes and sewers, electricity cables and gas pipelines, telephone lines and other information networks. This has lead to corresponding demands for goods and services. It is also reflected in an increased use frequency of household functions such as heating and washing.

23.4   Scope and Objectives of the HOMES Program

HOMES is aimed at developing and applying the concepts, operational approaches, methodologies, and instruments (e.g. models and scenarios) relevant for the diagnosis and evaluation of household metabolism in a complex western society. Contemporary household metabolism is not expected to be sustainable nor qualitatively acceptable. Therefore HOMES will also investigate the change necessary to accomplish a transition to household metabolism complying with these characteristics. Such a transition should result in a sustainable, adequate and equitable match between supply and demand of resources. A historical perspective is created by considering the situation throughout the 1950-1990 period, and for the years 2015 and 2050. The year 2015 is chosen because it is the time horizon of most long-range economic forecasts and because adequate details regarding cleaner production and consumption processes are available. The year 2050 is selected to study future conflicting demand and supply patterns concerning natural and energy resources: in the period until 2050 much tighter environmental quality standards can be expected to be implemented and accepted.

Three main kinds of household functions have been selected for detailed investigation. The first, from the category of infrastructure/housing, is heating and mobility. The second from the category of durable household goods, is white goods appliances. The third, from the category of non-durables, is consumption of water, detergents, gas and electricity. Together these three categories constitute the majority of the material and energy flows, including waste production and handling, that make up household metabolism.

The total demand for natural resources is determined by the number of households (and their average size) by the consumption per household, and by the material and energy efficiency of consumption. The latter is not only a function of spatial, bio-physical, technical, economic and behavioural aspects, but also of specific social institutions and administrative policy measures. Within the HOMES research team these various aspects are covered by different scientific disciplines, each of which investigates different aspects of household metabolism using specific sets of constraints. The different constraints put forward by the different disciplines ultimately determine the potential for change of household metabolism along sustainability and environmental quality lines as established in the evaluation phase of the program (cf. Figure 23.2).

The emphasis is first on metabolism of households in The Netherlands. In a later phase, the household metabolism methodology can be applied to other countries. The diagnostic phase has focused on a number of research questions, including the following: Which relevant trends can be observed regarding household metabolism in The Netherlands in the last decades and what are the impacts from household metabolism on environmental quality? To what extent are the observed trends caused by changes in the consumption per household and/or by changes in the energy and material intensity of consumption, due to e.g. technological developments, spatial, economic, behavioural and social changes or administrative measures? Finally, which conclusions about the future development of the use of resources for household consumption in The Netherlands can be derived from trend extrapolation?

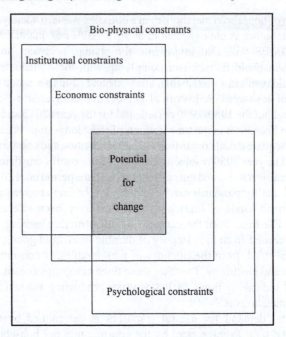

Figure 23.2   Different sets of constraints determine the potential for change of household metabolism

Source: Noorman et al. (1997)

23.5   Some Conclusions from the Diagnostic Phase

Spatial Perspective

From the beginning of the 20th century until 1970 dwellings in The Netherlands became increasingly larger and elaborately equipped, and were designed with functionally divided spaces. During the last two decades dwelling design developed towards a more neutral and flexible arrangement of space. Since the 1990s sustainability and environmental quality considerations have increasingly been taken into account in dwelling design. The indoor space per person has increased from 178 m^2 in 1950 to 225 m^2 in 1989, corresponding with declining average household size and increasing dwelling size. During the same time the average number of inhabitants per square kilometre increased considerably, from 309 in 1950, to 445 in 1994. This trend matches the fact that an increasing share of the population lives in urbanised areas. Since the share of people living in urban municipalities remained more or less constant at around 50 per cent, this trend can be well described as urbanisation of the rural area.

Urban Planning Perspective

A significant portion of the households is located in urban areas. The quality of the existing housing stock will strongly dominate the characteristics of household metabolism in the coming decades: 80 per cent of the projected housing stock in The Netherlands for 2015 already exists in 1995.

Because of the urbanisation rate and its associated generally negative environmental impacts the issue of urban environmental sustainability has become increasingly important. Urban production, consumption, and disposal are usually regarded as being detached from the overall ecological viability of the urban system, and are clearly different from nature's own metabolic processes.

Physical Perspective

Invariably the post-World War II household consumption rates of energy, water, durable consumer goods, and the generation of wastes related to the rapidly rising consumption patterns show rising trends. Energy is used for heating purposes, hot water production, cooking, transportation, lighting and an increasing number of electrical appliances. Electricity consumption increased by 9 per cent a year until the mid-1960s. After that the explosive growth of electricity consumption levelled off to an annual increase of about 2 per cent. Parallel to the rising number of cars (now on average one car per household), and distances travelled with these cars, fuel demand for passenger transport has increased five-fold since 1960.

House insulation measures reduced residential heat demand after the second oil crisis. Yet the period 1965-1995 is dominated by such factors stimulating energy demand as the discovery of the '*Slochteren*' natural gas field, investments in grids, decreasing energy prices, rising spending power, and the introduction of many electrical appliances. During this period the indirect energy consumption of Dutch households accounted for more than 50 per cent of the total primary energy requirements of households.

Tap water consumption of households almost tripled in the 1950-1992 period. Although the tap water consumption per household remained fairly constant, water consumption per person increased by 50 per cent. Parallel to the increasing consumption patterns, all forms of waste flows from households increased significantly. Since 1950 the amount of waste generated showed a four-fold increase, resulting in 384 kg waste per person in 1992. Households in cities produce now on average about 5 kg of waste less per week than households in rural areas (17 kg vs. 22 kg, respectively).

Psychological Perspective

Consumer behaviour was studied using the so-called MOA model. Motivation, Opportunity (external facilitating conditions), and Ability (internal capacity of consumers) are important factors determining consumption patterns. These three factors are related to other household characteristics such as age, household composition, and geographic location. In the MOA model, opportunity and ability are regarded as stimulating and limiting factors respectively, which influence the relation between consumer motivation and behaviour.

A field study was held among Dutch households. The first results indicate that income differences are strongly related to the possession and use of domestic appliances, and of the consumers' perception and evaluation of quality-of-life and environmental issues. Age turned out to be the second most important variable related to differences in perceptions and behaviour. It appeared that respondents with high-intensity consumption patterns (high income, middle-age, living outside cities) evaluate environmental issues more seriously than respondents with low-intensity consumption patterns.

Economic Perspective

Prices are regarded as a major policy instrument to regulate consumption. Analysis of past price developments of energy and water showed that the nominal energy prices and water prices increased due to an increase of VAT and other taxes, had the effect of regulating consumption. Furthermore, an extra tax was introduced to finance environmental activities. However, real energy prices dropped considerably during the 1950-1992 period.

Concerning the use of energy and water it can be concluded that prices, tariff structures, and the increasing efficiency of white good appliances did not stimulate energy and water saving behaviour. Price intervention was effective in reducing domestic waste flows. Fuel prices seem to influence mobility in two ways: the number of kilometres is reduced as fuel prices increase, and the decision to purchase or replace a car is postponed or cancelled, when fuel prices rise considerably.

Public Policy Perspective

Public policies strongly determine and influence household behaviour. During the last decades an increasing number of specific public policies have been implemented, aimed at adjusting environmental behaviour. Besides such specific policies, behaviour is also determined by policies that are not specifically intended to influence environmental behaviour. It appeared that in the past, behaviour in households was influenced mainly by policies that were not intended to control this behaviour (related to mobility, heating and white goods appliances). These non-specific policies mainly involved those whose side effects had a stimulating impact on these aspects of behaviour. Examples are public housing policy, physical planning, income policy and emancipation policy. The driving and counteracting forces which have been most influential in changing household metabolism rates and the related environmental impacts are summarised in Table 23.1.

The overall evaluation of outcomes of the diagnostic phase of HOMES shows that future demand for resources due to household consumption will exceed the potential of a sustainable and qualitatively acceptable supply of these resources. Therefore, 'sustainable options' which aim to decrease the household metabolism throughput have to be (re-)designed and ways of implementation need to be indicated. This will be studied in the *change* phase of HOMES. Field studies and scenario and model building will be used extensively during this phase.

Table 23.1   Factors underlying developments in household metabolism

Driving Forces	Counteracting Forces
• increasing population; increasing number of households	• technological innovations
• rising incomes; declining prices	• specific policy measures
• increasing opportunities and abilities	• increasing environmental awareness and attitudes
• non-specific policy measures	

23.6   Conclusions

HOMES adopts a metabolic view of goods and services. From this perspective urban environmental quality is to be seen in relation to its relevant attributes both inside and outside of cities. As demonstrated above, the distinction between direct and indirect aspects is especially relevant in this regard. The following examples will serve to illustrate this. In an all-electric clean city with electric cars and high quality public transportation the electricity needed may be generated outside the city from polluting fossil fuels. A nice city may well be built from sustainable materials taken from hitherto unspoilt countrysides or tropical forests. Bringing people and households closer together in compact cities may save energy for space heating but may increase demand for lighting. People living in compact cities may bicycle to work but may also be tempted to flee the city during the weekend or use air travel to go to a holiday destination. Compact cities are excellent locations for energy cogeneration plants but like any form of systems integration this has potential disadvantages in terms of increased risks from power failures etc. Households in cities may generate less direct waste on site but they may indirectly generate more waste elsewhere.

Summarising, compact cities may be beneficial for present local and regional environmental quality, but that quality should not be judged separately from environmental impacts elsewhere or later. Ultimately there is but one planet and its overall and global environmental quality counts now and in the future!

Notes

1   Prof. dr. A.J.M. Schoot Uiterkamp, K.J. Noorman and W. Biesiot are from the Department of Energy and Environment at the University of Groningen.
2   The contributions of the members of the HOMES team, A.G. Bus, A.M.J. van Diepen, B. Gatersleben, J.J. Ligteringen, V.G.M. Linderhof and J. v.d. Wal, are gratefully acknowledged. This research was partly sponsored by NWO, the Dutch Organisation for Scientific Research.

References

Diepen, A. van, and H. Voogd (1994) *Environmental quality and household behaviour*, Paper presented at the Regional IGU Congress at Prague.

Lansing, J.B., and R.W. Marans (1969) Evaluation of neighbourhood quality, *AIP Journal*, pp. 195-199.

Noorman, K.J., and A.J.M. Schoot Uiterkamp (1997) *Green households? Domestic Consumers, Environment and Sustainability*, Earthscan, London.

Opschoor, J.B. (ed.) (1992) *Environment, economy and sustainable development*, Wolters-Noordhof, Groningen.

Opschoor, J.B., and F. van der Ploeg (1990) Duurzaamheid en milieukwaliteit: hoofd-doelstellingen van milieubeleid. In: *Het milieu; denkbeelden voor de 21e eeuw*, [Environment: visions for the 21st century], Commissie Lange Termijn Milieubeleid. Kerkebosch, Zeist.

Pearce, D., A. Markandya and E.B. Barbier (1989) *Blueprint for a green economy*, Earthscan, London.

Pezzey, J. (1989) *Economic Analysis of Sustainable Growth and Sustainable Development, Working paper no. 15*, The World Bank, Environment Department, Washington, DC.

Poll, R. van (1997) *The Perceived Quality of the Urban Residential Environment*, Ph.D. Thesis, University of Groningen, Groningen.

Pugh, C. (ed.) (1996) *Sustainability, the Environment and Urbanization*, Earthscan, London.

VROM (1988-1989) Nationaal Milieubeleidsplan (NMP) [National Environmental Policy Plan (NEPP)], Tweede Kamer der Staten Generaal, SDU, The Hague.

WCED (1987) *Our common Future*, World Commission on Environment and Development, Oxford University Press, Oxford/New York.

WCN/IUCN (1980) *World Conservation Strategy: Living resource conservation for sustainable development*, World Conservation Strategy/International Union for the Conservation of Nature, Gland, Switzerland.

Chapter 24

Managing Information: Urban Environments and the Internet

H. Srinivas[1]

> If every human group had been left to climb upward by its own unaided efforts, progress would have been so slow that it is doubtful whether any society by now could have advanced beyond the level of the stone age... (Ralph Linton, 1934)

24.1 Introduction

Cities and towns in most countries around the world have been gaining considerable attention due to the large number of households migrating to cities and its consequent effects. It has also been due to the centrality of goods and services that cities offer. Over the last few decades, they have emerged as the major form of settlement. At the beginning of this century more people lived in and around cities than in rural areas. In 1800, only 50 million people lived in towns and cities worldwide. During 1975, there were 1.5 billion, and in 2000 there were three billion – more than the entire population of the world in 1960. (Megacities 2000, 1996: codex.html) Proximity to decision-makers and financial markets, large pools of skilled and unskilled workers, and other advantages have made such urban areas the engines of growth for the countries and regions where they are situated. For example, despite the environmental and social problems that it is facing, Bangkok's contribution to the national GDP has been estimated to be more than the combined output of all other cities in Thailand (ESCAP, 1991).

The result of this has been the explosive growth of urban areas, bringing with it a host of negative effects. Population concentration in increasingly smaller land masses has caused a drastic decline in the quality of living – both in the residential and work fronts. Cities have, in effect, become a barometer of humankind's progress into the 21st century, whether this is an upward trend or downward. Such a scenario has had ripple effects on a variety of sectors, such as education, health, labour/job markets, and economic activities, both directly and indirectly.

The growth and effect of an urban area should be seen not only in terms of its immediate boundaries, but also in terms of the resources necessary to sustain its population. A telling example is that of the Greater London area – the land mass that generates the resources necessary to sustain the population of Greater London, called the 'urban footprint', is actually a little more than the land area of UK! This illustrates the complex interrelationship and interdependence of urban areas and their surrounding hinterland. The effects of activities in urban areas (Table 24.1) have in many cases outweighed the relative agglomeration and centrality advantages that they offered. Thus, along with the benefits of urbanisation come environmental and social ills, including lack of access to drinking water and sanitation services, pollution and carbon emissions etc. A wide variety of urban problems can be observed, grouped under two broad classes, those associated with poverty and those associated with economic growth and affluence (WRI, 1996: ud_txt1.html).

Table 24.1   Components of the urban environment

Resources	Processes	Effects
Human Resources Sunlight Land Water Minerals Electricity Fuels Finance Intermediary products Recyclable materials	Manufacture Transportation Construction Migration Population Growth Residence/Living Community Services (Education, Health ...)	• Negative Effects: Pollution (air, water, noise), waste generation (garbage, sewage), congestion, overcrowding • Positive Effects: Product value-addition, increased knowledge base/ education, access to resources and better services

While the causes for these problems are many, focus has been maintained on the role and contribution of urban planning processes to this situation. The processes involved in urban planning and development vary considerably, and depend on a number of objective and subjective aspects in the physical, social, economic, and political spheres. In general, planning involves the cyclical processes of plan and policy-making, public debates and feedback, and its implementation and evaluation. A plurality of actors are involved in these processes, such as local governments, citizens groups, industry, governmental ministries, departments and other agencies, and the planners themselves. The interaction and intersections between these affect the overall development of the urban environment and the quality and attributes of the urban environment.

24.2   Policy Framework for Urban Environmental Management (UEM)

Historically, interaction between the various actors involved in UEM processes has been very weak and ineffective. While laws to effect such involvement existed, it was not exercised both on the part of local governments (adequate information was not provided), as well as other actors and citizens themselves (there was no commitment to participate). Information that was shared by the government was, in many cases, partial or selective. This put the entire decision-making process in the hands of the government as the main actor.

There has, however, been a growing awareness of environmental problems and its causes and effects. With a gradual increase in the transparency and openness in the functional organisation and operation of governments, legislation on information disclosure has been receiving considerable importance. Parallel to this has been a movement among the citizens to not only be aware of the processes of UEM within their community, but to also be involved in the design of decision making process itself. This calls for a major change in the basic understanding of citizens' participation and the consequent needs of information for decision making processes – from the points of view of all actors involved. With their direct involvement, the citizens of a community can be seen as major actors and partners in the process of planning. Such a give-and-take of information and decision support not only links the planning sector and the community, but also all sectors of the local government that affects the development of a region. Community involvement becomes all the more critical when the shortcomings and weaknesses of the local government to effectively deal with the range of problems are taken into account.

At the lowest level, community involvement can be seen as passive acceptance, where the community reorganises and adjusts to the implemented public plans. Public sector plans then become a base on which private decisions are made. At higher levels of participation, however, the community is directly involved in the decision-making process at all levels. Thus, matching and synchronising public plans to private/individual plans become important, where public services are developed so that the private/individual plans can function and be implemented efficiently. It also calls for open and free participation at all stage of the process and with no restrictions or barriers.

This is not to discount the roles of the government at the national, regional and local levels. A powerful argument remains for a strong government role in environmental management (Devas and Rakodi, 1993). Governments are needed to plan for growth, to regulate polluting activities, to harmonise competing uses of the urban environment, and to address questions of equity that purely market-oriented approaches do not cover. In efforts to improve the urban environment, local governments are especially critical since they are responsible for most aspects of environmental management at the city level, from the provision of urban infrastructure and land use planning to local economic development and pollution control.

Thus, interaction between the different actors at different levels of the planning processes and cycles becomes critical to respond to the increasingly complex policy and investment choices that urban communities face. There are many key points that arise in support of a sound urban environmental policy (IFEI, 1992). The concept of sustainable development ought to take into account the needs of future generations in decisions on how and whether to use resources and apply technologies, this is done through policies, investments and development plans. A clear position within a policy framework should be delineated, based on natural ecosystems at national, regional and local levels, as well as on human-environment interactions. It should take into account the stresses and effects occurring within these ecosystems. A system of prioritisation has to be established which incorporates human life, health, depletion and productivity of resource stocks, capacity of the environment, and systematic accounting measures. Environmental policies have to be based on an understanding of the causes of environmental degradation and of the environmental impacts and cost-effectiveness of solutions, as well as the uncertainties associated with it. Policies should also contribute to greater public understanding of environmental issues through more open access to information and decision-making process. Operationalising environmental policies, therefore, require the integration of many interrelated economic, environmental, social and cultural factors. A key policy input that arises out of, and facilitates, this understanding, is information.

24.3   Information as a Key Policy Input

The need for information that is timely, accurate and 'packaged' is an important input for policy formulation and decision-making. Data, statistics and other quantitative and qualitative valuations in a variety of formats constitute such information. This enables decision-makers to develop strategies for action, manage natural resources, prevent and control pollution, and evaluate progress made towards goals and targets. Why has such critical importance been assigned to the collection, analyses, processing and dissemination of information, with respect to UEM?

    In principle, there is a need to broaden and deepen the knowledge database of scientific and technical information concerning the links between economic activities and the environment. New data and knowledge on different aspects of urban environmental change needs to be collected. Such data has to be collected with its end-use in mind, providing the appropriate information at the appropriate level. Monitoring systems and information management technologies need to be extended and improved. Long-term monitoring systems, training, easily understood technologies of information gathering, effective use of existing data, and increased speed of data transmission are some of the ways in which this can be achieved. While accessibility and relevance of environmental reporting need to be widened, there is also a need for strengthening and expanding partnerships among institutions that produce, analyse and disseminate environmental information. Geographical Information Systems (GIS) need to be used for spatially segregated data. Levels of aggregation and coverage need to be linked to different levels of decision-making and stages of implementation.

    This also places emphasis on the methodology of packaging and reporting information. The aims and objectives, emphases, and language used significantly alters the quality and quantity of information being disseminated. The media used – reports, print and electronic media, internet, audio and video cassettes, conferences and workshops – facilitate better understanding of the issues involved, and assist in decision making.

    Criticality of information needs at the international, national, regional and local levels have to be balanced by the information collection processes that complement the need and packaging of information in terms of its collection, analysis, interpretation, and reporting. Thus, information is used for the detection of changes in the environment. This is important in the identification and recognition of environmental issues and problems. It provides a basis for evaluation and decision-making. It also facilitates monitoring of policy and programme performance in the implementation stages. In order for information to play its critical role in policy development, there is a need for accuracy and spread of issues covered, adequate historical and geographical coverage, and its comparability and consistency in collection.

Effectiveness in the implementation of a policy can also be attributed to effective dissemination of information. The aim of information dissemination is public education and enlightenment, consensus building, and the promotion of awareness. With this in mind, the quickness and periodicity of dissemination, along with its form, appropriateness and accessibility, has to be considered in dissemination.

24.4 An Information System for Urban Environmental Management

With the objective of collecting information that is of good quality, reliable, timely, relevant, and processed in order to facilitate the processes of urban environmental management, the following design principles need to be kept in mind when developing an appropriate information system (UN-DPCSD, 1996).

- *Decentralisation* – It is necessary to keep the information collection process decentralised in nature, and proximate to data collectors and users, since they best understand its use and limitations.
- *Responsibility* – Collectors of data should be responsible for its accuracy and appropriateness. Data should also have 'meta-data' such as date, origin and conditions for access, and responsible organisations.
- *Transparency* – Information should be freely available for all purposes and at all levels. All decision-makers should have access to the same information with highest standards of reliability.
- *Efficiency* – Data should only be collected once, avoiding unnecessary duplication, and simplifying reporting requirements. Data, once collected, should be readily and rapidly available to all users.
- *Economy* – Investment in the system should be done in scale to the use and analysis.

Providing information at the appropriate level, to the appropriate actors is key to an efficient information system (IFEI, 1992). For example, in a policy decision-making process, scientists and resource users, researchers etc. are the actors involved in problem identification. Here information is produced and analysed through observation, research-oriented data collection and scientific analysis. When it is necessary to generate awareness of the problem with the wider public and action is demanded, then environmental non-governmental organisations (NGOs), community groups, the media, etc. are involved. Information here is used to draw attention to the seriousness of the problem, in order to motivate action to find a remedy. Alternative actions are formulated by policy analysts from different disciplines who work for NGOs, industry, government and universities/research institutions. In this case, information is used to determine responsibilities for problem mitigation, costs and consequences of the problems, and developing alternative solutions. Politicians and their advisors as well as industry persons decide on the course of action to be taken – for which they need information to evaluate the different courses of action, based on their individual constituencies and responsibilities. Monitoring and evaluation is critical for feedback on the action. Here again, scientists, resource users, statisticians, economists, managers, NGOs etc. are involved in generating information to monitor environmental conditions and changes that occur in response to the chosen course of action.

Thus, environmental information is crucial to assess the impact of human activity on the environment. It helps in managing the natural and man-made resources in a sustainable way. Sound and sustainable decisions can be made by anticipating environmental degradation of resources and prevent costly remedial action. Progress towards achievement of development goals can also be made, assessing long-term effects of management interventions. Table 24.2 illustrates the relation between policy process and stages of data production.

Table 24.2   Environmental information in the policy and data production processes

Stages of Policy Process	*Stage of Data Production*		
	Monitoring and data collection	Processing and analysis	Dissemination
Identification and recognition of issues	Detection of changes in the environment	Transformation to concise information	Education/ enlightenment
Evaluation and decision-making	Provision of basis for evaluation and decisions	Identification of cause-effect, cost-benefit analysis	Consensus-building
Implementation	Monitoring of policy performance	Evaluation of policy performance	Promotion of public awareness

Source: Nishioka and Moriguchi (1991)

24.5   An Example from Japan: The Virtual Planning Lab

The intersection of information needs and the development of global networks of computers – internet, can be illustrated with an example from Japan, the Virtual Planning Lab (VPL).

During the 1990s, the internet, has gained considerable importance as a communicative and adaptive means of sharing and disseminating information on UEM issues. Many subject-specific groups have grown on the internet, rallying around shared expertise and ideals, by the enablement of new forms of communication. The most obvious example of these new computerised communications media is electronic mail, the world wide web and bulletin boards or newsgroups – which are in fact just the first generation of new forms of information and communications media. The digital media of computer networks, by virtue of their design and the technology upon which they function, are fundamentally different from the current mass media of television, radio, newspapers and magazines. Computer networks encourage the active participation of individuals rather than the passive non-participation induced by television etc. Applications such as electronic mailing lists, conferences, and bulletin boards, serve as a medium for group or many-to-many communication.

What do these fascinating capabilities of the internet portend for UEM? Urban planning as a science and a practice has itself been evolving over the years. We now have a situation, where the increasing importance of better qualities of living has transformed the citizen to an active actor in the process, highlighting more participation and decentralisation of activities, at the same time reflecting the global contribution of local planning decisions.

Existing efforts of city and regional governments in utilising the internet for urban planning and management, as well as for interaction between the various actors/levels involved in the planning process, have only been exploratory. An example of this initiative is the city of Blackburg in Virginia, USA (http://www.montgomery-floyd.lib.va.us/pub/locgovt/bburg/) where city officials used the internet extensively in communicating with the citizens by making the plans available over the internet. The citizens were surveyed for their opinion on the city's development process, and asked for their comments on the comprehensive development plan prepared by the city council. This was then ploughed back into improving the plan. The success of the Blackburg experience has prompted many cities in the USA and Europe to use the internet for their city's development and management.

In general, for environmental management, there have been essentially three very broad areas where the internet has proved viable: (a) query processing – answering questions, enquires and requests; (b) sharing of ideas and information about policies, programmes and projects; and (c) database development on a variety of subjects. These and other advantages that the internet offers for urban planning underpin the development of the 'Virtual Planning Lab' (VPL) at Tokyo Institute of Technology in Tokyo, Japan.

The information requirements that arise out of and for UEM processes are of particular interest of, and are the focus of, the VPL. Such information can be dynamically changing intelligence that allows appropriate and current decisions to be taken, or archived information against which comparative analysis etc. can be made. Involvement of all actors and intermediaries in providing the required information becomes important, besides packaging of information so that it is better understood and appreciated. Of particular importance for the VPL are electronic database resources that can be accessed to obtain information and data on a wide variety of categories, as well as interactive communication forums that facilitate information sharing and exchange between different users.

Thus the goal of the VPL is to create an environment for the exchange and archival of information on and for planning. Out of the need for information exchange comes the function of communication, and that of archives comes the database function. These two activities underpin the basic concept of the VPL, and create a feedback loop between the two, generating an effective knowledge base of planning and development (Figure 24.1). Interactive communicative forums involve means for decision support both at the personal and group levels. These include: (a) idea formulation or visioning at the personal level, where the involved actors individually develop ideas and visions/goals for the planning process; (b) interactive communication at the group level, where the ideas are further developed and enhanced by communication among the actors, as well as the various implementing agencies and users in the process; and (c) collaboration and assimilation of the above process into a complete workable plan and its implementation.

Each of these levels of support involves multi-user and multi-level interaction, with parallel processes of development. The VPL can therefore be seen as a hub (Figure 24.2) that facilitates intra communication not only within the local government (G to G), but also between the government and community and *vice versa* (G to C and C to G), and within the community itself (C to C). This can be seen as a progression that gradually develops to also encompass the region in which the community is positioned, and beyond.

Such decision support and communication require dynamic data generation, both numerical and textual, that can suit the differing user groups and needs. VPL's databases have to be structured with data stratified at various levels – for example, physical, economic, environment, political, etc. – each with its own defining parameters. New data matrices can be dynamically developed on-the-fly by combining these data sets at each level into new combinations that enables horizontal, vertical and cross-level comparisons. It not only allows for new data matrices to be developed, but also facilitates generation of innovative solutions. Creating an effective conceptual basis and information system within the VPL is very important. This development process is in fact cyclical, each feeding onto the next and therefore strengthening and widening its scope and reach.

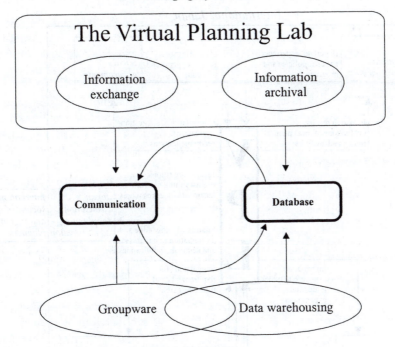

Knowledge base

System development includes:
- Survey of existing efforts in 'virtual planning'.
- Development and refinement of the VPL system (group collaboration for planning and personal information needs).
- Building up and designing the VPL system.
- Using the VPL system for 'experimental planning'.
- Evaluation of the system and the planning process.
- Reporting in the proceedings.
- Feedback and improvement of the system.

Figure 24.1   Basic functions and goals of the VPL

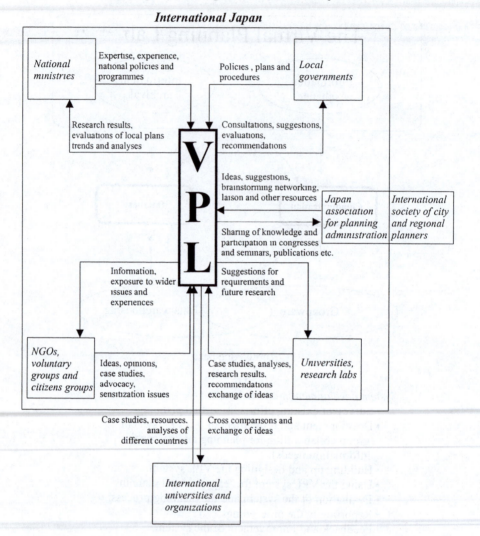

Figure 24.2   Operational framework of the VPL

The concept of the VPL rests on satisfying both personal information needs as well as facilitating collaborative group processes in the environmental management process. The alignment of the decision support system steps along a matrix is shown in Table 24.3.

Information sharing among the various actors is a sequential process, involving the disclosure of information, its exchange and interchange, and the convergence of options and opinions. These processes have to take place both within the organisations involved, as well as in interaction with other organisations. Such a matrix of decision support allows for vertical, horizontal and diagonal interactions, but more critically groups of cells can be supported in inter- and intra cycles, which results in a more consensual solution. The flexibility of the system is demonstrated by the fact that matrices can be used for each of the planning process stages. It is also significant that the information and processes output, which can be a plan, policy, project or programme, from each matrix can be input data for the next matrix (Figure 24.3). The VPL, as the name itself suggests, primarily uses the internet in order to facilitate the activities described above. Non-internet-based communication modes, including conventional and face-to-face, is also used for interaction.

Table 24.3   Uses of decision support sytem

	'SEE' Monitoring and Evaluation	**'PLAN'** Design and Plan-making	**'DO'** Implementation and conforming
Information Sharing	—	□	●
Decision Making	□	—	□
Acceptance	●	□	—

— Maximum criticality
□ Median criticality
● Minimum criticality

With the intention of developing a knowledge database of ideas the VPL's specific operations are essentially: (a) to collect plans, ideas, approaches, best practices/success stories from various local governments, research analyses and results/ recommendations from universities, research labs, and new technologies for a prudent and sustainable living environment for the future; (b) to process and analyse these collections to enhance their universal usability and applicability; and (c) to disseminate the information through the internet as well as face-to-face interactions. This can include publishing on the world wide web and print media, etc.

These activities are further supported by in-house and sponsored research projects, development and maintenance of communications systems, organising training, seminars and lectures etc. Such activities sustain interaction continuously on a variety of spheres and culminates in frequently organised national or regional intensive workshops which are life-long learning processes for local governments and its planners. Information on plans and programmes are submitted by local governments and then evaluated and analysed in these workshops, and suggestions for improvement are made. For example, local city governments present a problem or case study to a panel of experts appointed by the VPL, who then recommend corrective action and draw lessons. Such presentations and interactions take place on the internet, face-to-face, or a combination of both. Thus, workshops and meetings reinforce communication channels already established by telephone, fax and/or email (Figures 24.2 and 24.4).

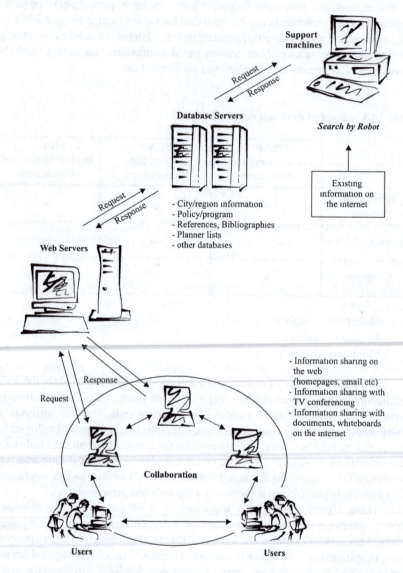

Support machines

Request
Response

Database Servers

Search by Robot

Existing information on the internet

Request
Response

- City/region information
- Policy/program
- References, Bibliographies
- Planner lists
- other databases

Web Servers

Response

Request

- Information sharing on the web
  (homepages, email etc)
- Information sharing with TV conferencing
- Information sharing with documents, whiteboards on the internet

Collaboration

Users

Users

Figure 24.3   Digital networks of the VPL

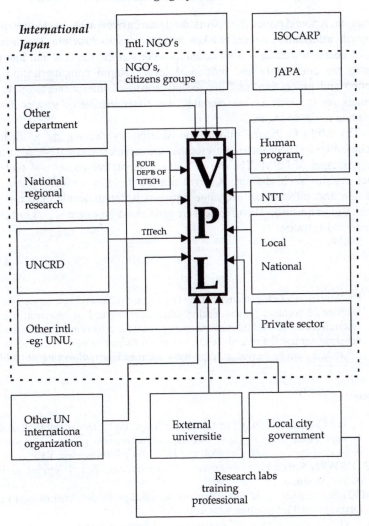

Figure 24.4 Organisational framework of the VPL

24.6 Conclusions

Chapter 40 of the Agenda 21 identifies two programme areas in order to emphasise the need that decisions are based on sound information. Clear organisational and operational delineations are outlined in the chapter to achieve this goal. The first area concerns a gap in currently available data that can be used for developing effective environmental measures, and the second concerns improving the availability of information. A sound policy framework for urban environments has to incorporate an appropriate information system as a key component in its structure. This involves all levels of decision making – from decision-makers at the national and international levels to the community and individual levels. Rapid computerisation of data collection and processing, and the recent growth of interconnected networks of computers on the internet has enabling the dissemination of greater amount of information to the end-user.

The VPL facilitates information sharing by serving as a medium for communication and decision support for the various actors and stake-holders involved in the processes of UEM. The use of various interactive means and tools on the internet enables such sharing. A level playing field is created, where smooth interaction and information exchange between different levels of actors can be effected. Commitment and involvement are generated by the ease of usage of computer hardware and software.

Note

1   Dr. Srinivas is a faculty member at the Department of Social Engineering at the Tokyo Institute of Technology, and teaches urban planning and development. His current research focus includes information management for urban environments, community development and the role of NGOs, and micro-finance in poverty aliviation. Further information can be viewed at: http://www.soc.ti-tech.ac.jp/titsoc/higuchi-lab/hari/.

References

Devas N., and C. Rakodi (1993) The Urban Challenge, in: N. Devas and C. Rakodi (eds), *Managing Fast Growing Cities: New approaches to urban planning and management in the developing world*, Longman and John Wiley & Sons, New York.

ESCAP (1991) *The State of the Environment*, UN Economic and Social Commission for Asia-Pacific, Bangkok.

IFEI (1992) *Proceedings of the Environmental Information Forum*, International Forum on Environmental Information, Montreal, May 21-24, 1991.

Linden, E. (1993) Megacities, *TIME International*, January 11, 1993.

Megacities 2000 (1996) Megacities Codex (http://www.megacities.nl/co-dex.html).

Nishioka, S, and Y. Moriguchi (1992) *Institutional Arrangement and Environmental Information Needs*, Center for Global Environment Research, National Institute of Environmental Studies, Tokyo.

UN-DPCSD (1996) *Report of the Workshop on Information for Sustainable Development and Earthwatch*, United Nations Department for Policy Coordination and Sustainable Development, Geneva.

UN Population Division (1995) *World Urbanization Prospects: The 1994 revision*, The United Nations, New York.

WRI (1996) 'World Resources: A Guide to the Global Environment, Special Issue on the Urban Environment' World Resources Institute (http://www.wri.org/wri/wr-96-97/ud_txt1.html).

Index